SpringerBriefs in Molecular Science

Green Chemistry for Sustainability

Series Editor

Sanjay K. Sharma, Jaipur, Rajasthan, India

SpringerBriefs in Green Chemistry for Sustainability present concise summaries of cutting-edge research and practical applications across a wide spectrum of fields including but not restricted to: Green House Gases and Global Warming, Microwave Synthesis, Solvent Free Reactions, Ultrasound Technology, Sonochemistry, Electrochemical Syntheses, Phytoremediation, Waste Management, Water Treatment and Wastewater Management, Dyes and pigments, Food and Pulp industry, Dairy Products and their Preservation, Catalysis and Biocatalysis, Biopolymer technology.

The publications are targeted at (but not restricted to) scientists and researchers in the fields of Chemical Engineering, Sonochemistry and Green Chemistry Research, Engineering and Innovative Technology, and Sustainable Development Issues, Environmental Pollution Abatement, Cleaner Production Technologies and Catalysis.

Featuring compact volumes of 50 to 125 pages, the series covers a range of content from professional to academic. Typical topics might include:

- A timely report of state-of-the-art analytical techniques
- A bridge between new research results, as published in journal articles, and a contextual literature review
- A snapshot of a hot or emerging topic
- An in-depth case study
- A presentation of core concepts that students must understand in order to make independent contributions
- Best practices or protocols to be followed

Briefs allow authors to present their ideas and readers to absorb them with minimal time investment. Briefs will be published as part of Springer's eBook collection, with millions of users worldwide. In addition, Briefs will be available for individual print and electronic purchase. Briefs are characterized by fast, global electronic dissemination, standard publishing contracts, easy-to-use manuscript preparation and formatting guidelines, and expedited production schedules. Both solicited and unsolicited manuscripts are considered for publication in this series.

Pratima Bajpai

Environmentally Benign Pulping

Pratima Bajpai
Consultant (Pulp and Paper)
Kanpur, India

ISSN 2191-5407   ISSN 2191-5415 (electronic)
SpringerBriefs in Molecular Science
ISSN 2212-9898   ISSN 2452-185X (electronic)
SpringerBriefs in Green Chemistry for Sustainability
ISBN 978-3-031-23692-1   ISBN 978-3-031-23693-8 (eBook)
https://doi.org/10.1007/978-3-031-23693-8

This Springer imprint is published by the registered company Springer Nature Switzerland AG
The registered company address is: Gewerbestrasse 11, 6330 Cham, Switzerland

Preface

This book provides professionals in the Pulp and Paper Industry with the most up-to-date and comprehensive information on the state-of-the-art techniques and aspects involved in environment-friendly pulping technologies. Traditional chemical and semi-chemical pulping processes are not environment friendly. So, it has become important to look for alternative eco-friendly approaches to mitigate wastewater emissions in paper industry, by making more stringent regulations to improve environmental conservation. In response to this problem, the industrial and scientific sectors are increasingly aiming at using new raw materials to replace traditional choices and also at developing new pulping processes based on less polluting, more easily recovered reagents. This book provides up-to-date, comprehensive insights into new and emerging deep eutectic solvents (DES) for lignocellulosic biomass pretreatment and discusses the effects of DES on biomass pretreatment and the production of value-added products. It also discusses biotechnological methods of pulping. Biotechnological processes help to make manufacturing processes cleaner and more efficient by reducing toxic chemical pollution and greenhouse gas emissions. This book will be an invaluable reference source for Applied Chemists, Foresters, Chemical Engineers, Wood Scientists, and Pulp and Paper Technologists/Engineers, Researchers and Experts, and anyone else interested or involved in the Pulp and Paper Industry.

Kanpur, India   Pratima Bajpai

Contents

List of Figures

List of Tables

Chapter 1
General Background and Introduction

Abstract One of the world's most important industries is the pulp and paper industry. More than 400 million tons of paper products are produced annually, indicating the significant global demand. Due to the industry's high capital requirements and low profit margins, new technology implementation in mills is typically restricted. As the industry is constantly under pressure to reduce environmental emissions to the air and water, an increase in production and improved environmental performance are required to meet the rising demand. Today's pulp and paper businesses are highly motivated to operate sustainably, and there is a growing demand to implement waste minimization strategies in order to establish a mill with the lowest environmental impact. There are a number of ways that the paper pulping process in facilities all over the world can become environmentally friendly if a sustainable development strategy is implemented.

Keywords Pulp and paper industry · Pulping · Kraft process · Sulfite process · Black liquor · Sustainability · Environmental emissions

More than 5 billion people in the world receive paper from the pulp and paper industry. Initially, pulp and papermaking required a lot of time and labour, but today's processes are fueled by expensive machinery and fast, high-tech paper machines. *Paper* and wood *products* are an *essential* part of daily life because they have promoted literacy and cultural advancement. Around 401 million metric tonnes of paper and paperboard were produced worldwide in 2020. (https://www.statista.com/statistics/270 314/global-paper-and-cardboard-production/). Graphic paper accounted for one third of production, and packaging paper accounted for more than half of that production. When compared to the prior year, there was a decrease of almost 1%. Since 2010, the production of paper and paperboard has averaged about 400 million metric tonnes annually, remaining largely stable. Atmospheric carbon dioxide levels were over 400 ppm at that point, reaching their most toxic levels.

China, the US, and Japan are the three nations that produce the most paper worldwide. Half of the world's total paper production is produced in these nations. The most important importers and exporters of paper are Germany and the USA. About 10.68 million tonnes of processed paper and cardboard were produced in China in April 2018.

China, the USA, and Japan are the three nations that produce the most paper worldwide. Half of the world's total paper production is produced in these nations. The top two countries importing and exporting paper are Germany and the United States. About 10.68 million tonnes of processed paper and cardboard were produced in China in April 2018. The global pulp and paper industry has declined slightly over the past five years due to the shift to digital media and paperless communications in most developed countries. A manufacturing boom in some emerging markets spurred increased demand for paper used in packaging, partially offsetting this decline.

Two of the industry's most promising growth areas, packaging materials and hygiene products, are now receiving more attention. Over the next five years, industry revenue is anticipated to slowly resume growing, but growth in developing markets will outpace growth in Europe and the US. $1270 billion USD has been calculated as the market size for paper and printing worldwide. In 2015 compared to 2014, it decreased by 6%.

Paper for writing and paperboard for packaging are just a couple of the many uses for paper products. The pulp and paper industry is energy intensive but less carbon intensive as biomass, which is considered carbon neutral, dominates the fuel mix. But in addition to ongoing fuel switching and breakthrough technologies, increased adoption of BAT and energy efficiency measures is necessary if the sector is for reaching its long-term decarbonization objectives.

Three types of significant environmental pollution are produced during the pulping stage of the paper manufacturing process. These issues include deforestation, water pollution, and air pollution. In the last 50 years, the use of paper has increased four times, according to recent reports on the global paper industry.

The pulping process is the most complicated one in the paper industry because it requires sulphur compounds to process the wood into pulp. Sulfur compounds frequently escape from the pulping process used to make paper and are released into the atmosphere. Sulfur compounds that are released during the paper-pulping process, such as hydrogen sulphide and dimethyl sulphide, have a strong odour. Due to the lack of odour control equipment in paper and pulp mills, these harmful emissions are released into the atmosphere.

In addition to posing environmental risks, the pollutant dimethyl sulphide has the potential to have an impact on the climate. Additionally, it alters the earth's energy balance by adding more aerosols to the atmosphere through condensation or nucleation on other areosols. This is apparent from the way sunlight reflects off of the earth directly, or indirectly through aerosols or changing clouds. In such circumstances, it is likely that the sun, a pure source of vitamin D, will become contaminated. Industries that manufacture paper emit nitrogen dioxide (NO), sulphur dioxide (SO), and carbon dioxide (CO). Acid rain is caused by all the pollutants released during paper production, and CO is a significant greenhouse gas that contributes to climate change. These harmful gases add to air pollution. Organochlorine compounds may also contaminate waste water.

These pollutants, which are highly flammable and cause health risks like eye irritation and skin infections, are released during the pulping process in the paper-making industries. In addition, the most recent United Nations report reveals that greenhouse gas emissions have increased over the past ten years. To achieve the 1.5 °C goal of the Paris Agreement, climate ambition must increase by at least five times. The world could soon experience a temperature increase of more than 3 °C.

The provisions of two major international environmental law declarations are clearly violated by environmental hazards caused by the aforementioned factors (IEL). This includes The Rio Declaration and the 1972 Stockholm Declaration of the United Nations Conference on the Human Environment (UN Doc. A/CONF/48/14/REV.1).

Environmentally unfriendly are conventional chemical and semi-chemical pulping processes. By enacting stricter regulations to improve environmental conservation, it is now crucial to look for alternative eco-friendly approaches to mitigate wastewater emissions in the paper industry. The industrial and scientific sectors are increasingly working to find new raw materials to replace conventional options as well as to create new pulping processes based on less polluting, more readily recoverable reagents as a solution to this issue.

The key initiatives taken by the pulp and paper industries to minimize pollution at various levels are mentioned below:

- Superior operating procedure
- Process modification
- Process redesign
- Recycling.

By adopting a sustainable development strategy, there are several ways that the paper pulping process in facilities all over the world can become an environmentally friendly phenomenon. These include installing environmentally friendly systems inside industrial buildings, minimising process modifications, and improving the operating conditions.

The scientific community started coming up with solutions to the numerous problems with the traditional processes in the 1970s, including the offensive smells, low yields, high pollution, difficult brightening of pulp, expensive investments, and significant energy, water, raw material, and reagent consumption. Initially, efforts were made to alter the pulping process; later, new methods without sulphur as a reagent were created. But these efforts failed. New issues like the challenging reagent recovery and the polluting nature of the waste emerged. Organic solvent-based new processes first began to appear in the 1980s. Their biggest benefit a full use of the raw materials was permitted. A few were employed to produce hydrolysable cellulose, sugars, and phenolic lignin polymers (Cox and Worster 1971; Muurines 2000; Oliet 1999; Varshney and Patel 1989).

The Kraft or sulphate process is the pulping method that businesses use the most frequently (FAO 2012). Raw materials used are wood and nonwoods. Wood is the most frequently used raw material in this process. The drawback of nonwoods is that some of them contain a lot of ash, which seriously impairs the black liquor recovery systems. Although the Kraft process continued to be used in the 1990s, the environmental issues it raised and the substantial investments required made it clear that alternate methods of producing pulp should be developed.

Sodium sulphate is the reagent that is replaced in this process, but the sulphur that is generated during the delignification reaction is the actual agent at work. The production of the pulp and the recovery of the used chemicals are the two steps that make up the process, respectively (Libby 1997).

Some of the black liquor can be recycled during Kraft pulping and used as a pulping liquor.

Black liquor makes up 40 to 60% of the pulping liquor in some circumstances without having an impact on the pulp yield or the properties of the pulps produced. In this manner, the penetration of the reagent into the chips is encouraged, a portion of them is reused without an expensive evaporation stage, and the heat energy of the black liquor is utilised.

The Kraft method has been modified by several authors. Anthraquinone is suggested by Wang et al. (2004) to be added to green liquor in order to increase yields by 2% while also making significant reagent and energy savings of 23 to 26%.

Luthe et al. (2004) reported that the pulp yield improved by 1.5 to 3.5% when polysulfides are used in the pretreatment.

Gustafsson et al. (2004) suggested pretreatment with polysulfide in an alkaline medium (0–2.5 molar sodium hydroxide) for improving pulp viscosity substantially while maintaining a lower Kappa value.

Hyperalkaline pulping with polysulfide involves two stages of pretreatment: an acid neutralisation stage and a high concentration of alkali and polysulfide stage (Brannvall et al. 2003).

The focus of subsequent research was on methods utilising organic solvents to separate not only cellulose fibre but also other useful components from the raw materials. By drawing an analogy between this and the fractionation of crude oil, the term "wood refinery" was created (Judt 1990). Numerous studies have been done on the delignification of conventional and non-conventional raw materials using organic solvents to produce pulp, lignin, sugars, and other products (Muurines 2000; Oliet 1999; Jiménez et al. 1997; Hergert 1998; Rodríguez et al. 1996, 1998; Neves and Neves 1998; Abad et al. 1997; Montane et al. 1998; Botello et al. 1999a, b; Lehnen et al. 2001; Guha et al. 1987; Bajpai 2021).

Comprehensive insights into new and emerging technologies for environmentally benign pulping technologies are discussed in this book. Biotechnological methods of pulping are also dealt with. By reducing emissions of greenhouse gases and toxic chemicals, biotechnological processes contribute to cleaner and more effective manufacturing processes.

References

Abad S, Alonso JL, Santos V, Parajó JC (1997) Furfural from wood in catalyzed acetic-acid media. Math Assess Bioresour Technol 62(3):115–122

Bajpai P (2021) Nonwood plant fibers for pulp and paper, chapter 7 pulping properties/pulping. Elsevier

Botello JL, Gilarranz MA, Rodríguez F, Oliet M (1999a) Recovery of solvent and by-products from organosolv black liquor. Sep Sci Technol 34(12):2431–2445

Botello JL, Gilarranz MA, Rodríguez F, Oliet M (1999b) Preliminary study on products distribution in alcohol pulping of eucalyptus globules. J Chem Technol Biotechnol 74(2):141–148

Brannvall E, Gustafsson R, Teder A (2003) Properties of hyperalkaline polysulphide pulps. Nord Pulp Pap Res J 18:436–440

Cox LA, Worster HL (1971) An assessment of some sulphur-free chemical pulping process. Tappi J 54(11):1890–1892

FAO Pulp and Paper Capacities 2011–2016 (2012) Available from http://www.fao.org/docrep/016/i3005t/i3005t.pdf

Guha SRD, Sirmokadan NN, Goudar TR, Rohini D (1987) Pulping rice straw by pollution free organosolv process. Indian Pulp Paper Tech Assoc 24(1):24–26

Gustafsson R, Freysoldt J, Teder A (2004) Influence of the alkalinity in polysulphide pretreatment on results of cooking with normal liquor-to-wood ratios. Paperi Ja Puu-Paper Timber 86:169–173

Hergert HL (1998) Developments in organosolv pulping. An overview. In: Young RA, Akhtar M (eds) Environmental friendly technologies for the pulp and paper industry. John Wiley and Soong Inc., New York, p 1998

Jiménez L, Maestre F, Pérez I (1997) Disolventes orgánicos para la obtención de pastas con celulosa. Review Afinidad 44(467):45–50

Judt M (1990) Recent developments in pulp production and their suitability for use in industrial process in developing countries. Cellul Sources Exploit 8:81–88

Lehnen R, Saake B, Nimz NH (2001) Furfural and hydroxymethylfurfural as by-products of formacell pulping. Holzforschung 55(2):199–204

Libby CE (1997) Ciencia, tecnología sobre pulpa y papel. México: CECSA

Luthe C, Berry R, Li J (2004) Polysulphide for yield enhancement in sawdust pulping: does it work? Laboratory simulations suggest that PS improves yield. Canada: Pulp Paper 105:32–37

Montane D, Farriol X, Salvado J, Jollez P, Chornet E (1998) Fractionation of wheat straw by steam-explosion pretreatment and alkali delignification. Cellulose pulp and byproducts from hemicellulose and lignin. J Wood Chem Technol 18(2):171–191

Muurines E (2000) Organosolv pulping. A review and distillation study related to peroxyacid pulping. [Thesis Doctoral]. Finlandia: Departamento de Ingeniería de Procesos, Universidad de Oulu

Neves FL, Neves JM (1998) Organosolv pulping: a review. Papel 559(8):48–52

Oliet M (1999) Estudio sobre la deslignificación de Eucalyptus globulus con etanol/agua como medio de cocción. [Thesis Doctoral]. Madrid, Spain: Departamento de Ingeniería Química, Universidad Complutense de Madrid

Rodríguez F, Gilarranz MA, Oliet M, Tijero J, Barbadillo P (1996) Manufacture of cellulose pulps by organosolv processes. Investigación y Técnica Del Papel 33(130):839–857

Rodríguez F, Gilarranz MA, Oliet M, Tijero J (1998) Pulping of lignocellulosics by organosolv processes. Recent Res Dev Chem Eng 2:9–47

Varshney AK, Patel DP (1989) Biomass delignification-organosolv approach. J Sci Ind Res 47(6):315–319

Wang SF, Ban WP, Lucia LA (2004) The effect of green liquor/anthraquinone-modified Kraft pulping on the physical and chemical properties of hardwoods. Appita J 57:475–480

Chapter 2
Pulp and Paper Making Processes

Abstract Raw material preparation and handling, pulp manufacturing, pulp washing and screening, chemical recovery, bleaching, stock preparation, and papermaking are the stages of the papermaking process. The primary two steps in the production of paper are the transformation of a fibrous raw material into pulp and the papermaking of the pulp. First, the harvested wood is processed to separate the fibers from the lignin, the unusable part of the wood. Chemical or mechanical processes can be used to produce pulp. Depending on the kind and grade of paper that will be produced, the pulp is then bleached and further processed. The pulp is dried and pressed in the paper factory to make paper sheets. A growing proportion of paper and paper products are recycled after use. Paper that is not recycled is either thrown away or burned.

Keywords Pulp and paper making · Raw material preparation · Pulp manufacturing · Pulp washing · Screening · Chemical recovery · Bleaching · Stock preparation · Papermaking

The pulp and paper sector has a wide range of products. Different types of raw materials are used to create a wide range of paper types using various processes in mills of different sizes. Pulp and paper are produced using cellulose-containing raw materials, most often wood, recycled paper, and agricultural waste. About 60% of cellulose fibres in developing nations come from non-woody raw materials like bagasse, bamboo, esparto grass, reeds, flax, jute, cereal straw, sisal, and plant fibres (Bajpai 2018a).

The production of paper goes through many stages. This includes: preparation of the raw materials, pulping, washing and screening of pulp, chemical recovery, bleaching, preparing stocks and making paper (Table 2.1). Basically, paper is produced in two steps, raw material is first converted into pulp, and then into paper (Gullichsen 2000; Bajpai 2015; EPA 2001b).

The raw material is first treated to separate the fibres from the lignin, which is the unusable part of the wood. Both mechanical and chemical methods can be used to make pulp. The pulp is then bleached and put through additional processing depending upon the type of paper to be produced. After bleaching, the pulp is subjected to drying and pressed to create paper sheets in the paper factory. An

Table 2.1 Steps involved in the manufacturing of pulp and paper

Raw material preparation
Debarking
Chipping and conveying
Pulping
Chemical pulping
Semichemical pulping
Mechanical pulping
Recycled paper pulping
Chemical recovery Evaporation
Recovery Boiler
Recausticizing
Calcining
Bleaching
Mechanical pulp bleaching
Chemical pulp bleaching
Stock preparation and papermaking
Preparation of stock
Dewatering
Pressing and drying
Finishing

increasing percentage of paper products are recycled after use. Paper that cannot be recycled is burned or dumped on the ground. Paper mills and pulp mills can operate separately or together. Pulp is used as a source of cellulose for the production of fibre and paper even cardboard.

2.1   Raw Material Preparation

Although other raw materials may be used, wood is mostly used for making pulp. When using wood as a raw material, the first steps in the pulping process are debarking, chipping, chip sorting, handling and storage, and other processes like depithing (for instance, when bagasse is used). After entering the pulp mill typically as logs or chips, the wood is processed in the woodyard. Operations in the woodyard are typically unaffected by the kind of pulping method used. If the wood is brought into the woodyard as logs, several processes transform the logs into chips suitable for making pulp (Ressel 2006; Bajpai 2018a; Gullichsen 2000; Smook 1992).

The slasher cuts the log to the desired length. This is followed by debarking, chipping, chip sorting and transportation to storage. Wood chips are typically stored on site in bulk. Chips are screened, washed and stored for further processing for a short period. Certain types of mechanical pulping processes, like stone groundwood pulping use roundwood; but most of the pulp production requires wood chips. Chips of uniform size are required for process efficiency and pulp quality. Then the chips are sent through a series of vibrating screens for removing oversized or undersized chips. Larger chips remain in the top screen and are sent for re-slicing, whereas smaller chips can usually be burnt with the bark or sold for other uses. It is processed according to its composition to minimize fiber degradation and maximize the pulp yield. Generally non-woody raw materials are stored in bales. The major products in the debarking process are wood chips. Bark, is the by-product. The bark is used as fuel or sold outside for other applications. The bark is commonly used for power generation. It is used as fuel in burners (Ressel 2006).

The logs are converted to chips for the ensuing pulping processes after debarking. The most popular type of chipper is a flywheel type disc having a number of knives mounted radially along its face. Substandard chips negatively affect pulp processing and the quality. Transferring the chips through multiple stages of vibrating sieves typically separates acceptable size chips from fine and oversized debris. The excessively large chips are rejected and transported to a "rechipper" by a conveyor. Typically, the fines and bark are burned together (if dedicated pulping equipment is not available). The majority of chips are moved within mill areas using an airveying system, either on belts or in pipes. It is common to use chip storage because chips are easier and less expensive to handle than logs.

To prevent the contamination of fresh chips by old chips, chips should be stored in a first-in, first-out order; the ring-shaped pile makes it easier to completely separate "old" and "new" chips. For a given wood source, size uniformity (that is, length and thickness) and 'impurity' content are considered indicators of chip quality. All chips that are 10 to 30 mm long and 2 to 5 mm thick are typically regarded as being of high quality. Oversized chips, pin chips, fines, bark, rotten wood, dirt and other materials are all regarded as contaminants. In chemical pulping, oversized chips are the primary source of screen rejects and are a handling issue.

It is challenging to reduce the size of the oversize fraction without imposing fines. When cooking chemical pulps, pin chips, particularly fines, and rotten wood contribute to issues with liquor circulation and reduced yield and strength in the resulting pulps. Particularly in mechanical and sulfite pulping, bark primarily represents a dirt issue. The majority of bark particles are soluble in the alkaline liquor, making the kraft process much more tolerant to bark.

2.2 Pulping

Pulp is produced by using mechanical (including thermomechanical), chemimechanical, and chemical pulping processes (Gullichsen 2000). Table 2.2 shows the main types of pulping processes.

Table 2.2 Types of pulping processes

Pulp grades
Chemical pulps
Sulfite pulp
Softwoods and hardwoods
Fine and printing papers
Kraft sulfate pulp
Softwoods and hardwoods
Bleached-printing and writing papers, paperboard, Unbleached-heavy packaging papers, paperboard
Dissolving pulp
Softwoods and hardwoods
Viscose rayon, cellophane, acetate fibers, and film
Semichemical pulps
Cold-caustic process
Softwoods and hardwoods
Newsprint and groundwood printing papers
Neutral sulfite process
Hardwoods
Newsprint and groundwood printing papers
Mechanical pulps
Stone groundwood
Mainly Softwoods
Corrugating medium
Refiner mechanical (RMP)
Mainly softwooods
Newsprint and groundwood printing papers
Thermomechanical (TMP)
Mainly softwooods
Newsprint and groundwood printing papers
Chemi-mechanical (CTMP)
Mainly softwooods
Newsprint, Fine papers

2.2.1 Mechanical Pulping

When mechanical energy is applied on the wood during mechanical pulping, the bonds holding the fibres together slowly get broken, releasing fibre bundles, single fibres, and fibre fragments. The advantageous printing properties of mechanical pulp are a result of the fibre and fibre fragment mixture. The goal of mechanical pulping is to preserve the majority of the lignin for producing a higher yield with acceptable brightness and strength characteristics. Mechanical pulps tend to discolour because these pulps show lower resistance to ageing (Bajpai 2018a; Gullichsen 2000; Smook 1992).

The most important mechanical pulping processes are stone groundwood pulping, pressure groundwood pulping, thermomechanical pulping or chemithermomechanical pulping. Wood is ground into pulp during the groundwood pulping process. In order to remove small pieces, a log is typically pressed against a rotating surface. After cooking, the groundwood pulp is frequently softened. Newsprint is made from this pulp. It is also used in lower-cost book grades. It increases opacity, bulk and compressibility. Given that all of the wood is used, groundwood pulp is inexpensive; however, contains impurities that may cause the paper to become discoloured and weak (Arppe 2001).

In chemimechanical process, mechanical abrasion as well as chemicals are used. Thermomechanical pulps, which are employed in the production of goods like newsprint is produced from raw materials using heat and mechanical operation. In this process high temperature steaming prior to refining is used; this softens the lignin between the fibers and partially removes the fibers. It is the outer layer of the fibers and exposes the cellulose surface for interfiber bonding. TMP pulp is usually stronger in comparison to groundwood pulp. This allows a lower furnish of reinforcing chemical pulp for magazines and newsprint. It is also used as a raw material for printing paper, tissue paper and cardboard. Softwood is mostly used for producing TMP. Strength properties are inferior in case of hardwood pulps. These fibers do not produce fibrils when refined. However, these get separated into shorter stiff debris. Therefore, TMP hardwood pulp is mainly used as a filler pulp. These pulps exhibit high purity and high scattering coefficients. Chemimechanical pulping and chemithermomechanical pulping (CTMP) processes are alike; use lesser mechanical energy and softens the fibers with sodium sulfite, sodium carbonate or sodium hydroxide. CTMP pulp is also good when using hardwoods, as long as the reaction conditions are suitable to yield a higher level of sulfonation. As compared to chemical pulp, mechanical pulp is weaker but can be produced at a lesser cost (about 50% of the pulp cost).The yield typically ranges from 85 to 95%. Mechanical pulp now accounts for about 20% of total virgin textile material.

2.2.2 Chemical Pulping

Chemical pulp is produced by cooking the raw material using the kraft (sulfate) process and the sulfite process. Chemical pulps are used in most commercially produced papers in the world today.

Kraft Process

A variety of pulps are produced by the kraft process and are primarily used for producing packaging, high strength paper, and board. Sodium hydroxide and sodium sulphide are the active cooking agents (white liquor) used in the kraft pulping process. All types of wood species can be processed using the Kraft process, but due to its chemical makeup, it has the potential to produce compounds with a bad smell. Kraft pulp shows stronger pulp properties as compared to sulfite pulp. The superior pulp strength of the kraft process and the advantages in chemical recovery have made it the industry standard. It accounts for 75% of all produced pulp and 91% of chemical pulping. Several grades of pulp are generally produced and the yield varies depending on the quality of the product.

Dark brown unbleached pulps are typically used for packaging products. This is because it is cooked to a high yield and retains most of the original lignin. White papers are produced from bleached pulp grades. The grades with the lowest yields, bleached grades, account for nearly half of kraft production. Since the early 1980s, when modified cooking technology was introduced, the advantages of kraft pulping have grown even more. Through ongoing research and development, different types of kraft pulping processes have been developed. Examples of continuous cooking are modified continuous cooking, isothermal cooking, and compact cooking. Examples of batch cooking technology are cold blow, SuperBatch/rapid displacement heating, and continuous batch cooking (Bajpai 2018a; Gullichsen 2000; Smook 1992; Sixta 2006).

Sulfite Process

Different chemicals are attacked and removed from lignin using the sulfite process. When compared to the kraft method, which can only use highly alkaline broth, sulfite process shows a high degree of flexibility. By varying the chemical dosage and makeup, the entire pH range can be used for sulfite digestion. As a result, a wide variety of pulp types and qualities can be produced using sulfite pulping for a variety of applications. Acid bisulfite, bisulfite, neutral sulfite (NSSC), and alkaline sulfite are the four primary sulfite pulping methods. The sulfite pulping method that is most widely used in Europe is the *magnesium sulphite pulping. Few mills are using sodium as base.* Chemical recovery is possible with magnesium as well as sodium bases. Lignosulfonates produced from pulping liquor can be utilized as a starting material for various chemical products (Smook 1992).

2.2.3 Semi-chemical Pulping

This process combines chemical and mechanical energy for extracting the pulp fibers. Inside the pulper, the wood chips are partially softened by chemicals, steam and heat. After this, the pulping process is terminated by mechanical means. The chemicals and organic compounds dissolved in the pulping liquor are removed from the pulp by washing. To improve machinability, the virgin pulp is blended with 20 to 35% recycled fibre (for instance, double-lined kraft corrugated clippings) or repulped secondary fibre (for instance, used corrugated containers). The semichemical pulping process in current operating mills use the sulfur-free or neutral sulfite semichemical (NSSC) process.

Sodium-based sulphite cooking liquor is used in the NSSC process whereas in the nonsulfur process sodium carbonate is used alone or combination of sodium carbonate and sodium hydroxide is used (EPA 2001a). The yield of semichemical pulps from hardwoods is in the range of 65 to 85%. The NSSC process is the most significant semichemical method. It uses a buffered sodium sulphite solution to partially pulp chips before treating them in disc refiners for complete separation of fibres. The lignin in the middle lamella is sulfonated and partially dissolved to weaken the fibers and prepare them for ensuing mechanical defibration. NSSC pulp is mostly used in the unbleached products which require high strength and high stiffness. Examples are corrugated paper, parchment paper, and bond paper. NSSC pulping is generally incorporated into Kraft mills for facilitating chemical recovery in a process known as cross recovery, which involves treating spent sulfite liquor with Kraft liquor. Spent sulfite solution provides the supplement needed for the kraft process.

2.2.4 Secondary Fibre Pulping

Waste paper is becoming an increasingly important source of paper fiber. Presently, recycled fiber accounts for about 50% of paper raw materials (Bajpai 2013). Recycled paper or cardboard is re-moistened and shredded into pulp during the recycling process, primarily by mechanical means. Chemical deinking and mechanical separation are the two methods for ink removal, adhesives, and other contaminants. Because of using closed water cycles and small-scale aerobic or anaerobic treatment system for the removal of some dissolved organic material from the recycled water, plants using non-deinked recycled paper can sometimes operate without waste water discharge. A closed loop makes sense if the product is able to tolerate a certain level of dirt and contamination. In some recycling plants, approximately 30 to 40% of the raw material processed ends up as sludge and must be disposed of as solid waste. For uses that don't call for a high level of brightness, processing recovered paper without deinking is adequate. To make the pulp lighter and cleaner, deinking processes are used for removing ink from the pulp. After deinking, bleaching is occasionally used as well. Recycled fibers are deinked for applications requiring higher brightness. Process water is comparable to water from undeinked systems. However, deinking reduces yields and necessitates additional internal processing. As a result, 30 to 40% of the material entering the process can end up in white water, which must be treated and removed before waste water can be discharged. The pulp yield can be only 60 to 70% of the waste paper that enters the process.

2.3 Chemical Recovery

Chemical recovery procedures are used by chemical and semichemical pulp mills to recover used pulping chemicals from the cooking liquor. Waste cooking liquor from the brownstock washers is directed to the chemical recovery section at soda and Kraft pulp mills. This process involves concentration of dilute black liquor, combustion of organics, reduction of inorganics, and reconstitution of pulping liquor (Vakkilainen 2000; Bajpai 2008). A dilute mixture of lignin, organic matter, oxidized inorganic compounds and white liquor forms residual weak black liquor from the cooking process. To produce thick black liquor, the thin black liquor is passed through multi-effect evaporators for increasing the solids content to approximately 50%. It is either sent directly to a non-direct contact evaporator, or concentrated further in a direct contact evaporator. In the direct contact evaporator, the odorous, fully reduced sulfur compounds are removed from the black liquor as it contacts the hot flue gases from the recovery furnace. It is possible to reduce these emissions by oxidising the black liquor before evaporation to prevent this from happening. Following the final evaporator/concentrator, the black liquor's solids content typically ranges between 65 and 68%. The soda recovery process is similar to the Kraft process but it is a sulfur-free process. It does not produce fully reduced sulfur, so in case of soda process black liquor oxidation system is not required (Adams 1992; Arpalahti et al. 2000).

The recovery furnace is then sprayed with the concentrated black liquor, where the organics are burnt; the sodium sulphate is converted to sodium sulphide. The higher energy content of the black liquor burnt in the recovery furnaces is recovered as steam for certain needs like heating and evaporation of the black liquor, preheating the combustion air, and drying the pulp and paper.

Power boilers that burn fossil fuels or wood are frequently used to supplement the process steam from the recovery furnace. Particulate matter that exits the furnace along with the hot flue gases is gathered in an ESP. This is mixed with the black liquor that is burned in the recovery furnace. The black liquor may also be given more makeup sodium sulphate.

Inorganic salts that have "smelted" or melted together gather at the furnace's base in a char bed. In the smelt dissolving tank, smelt is taken out and dissolved in diluted wash water to create "green liquor". This contains carbonate salts primarily composed of sodium sulphide and sodium carbonate. Additionally, dregs are present in green liquor. These are unburned insoluble carbon and inorganic impurities. These impurities are eliminated in the clarification tanks. Green liquor that has been decanted is moved to the causticizing area, where lime is added to convert sodium carbonate to sodium hydroxide. The lime kiln's calcium oxide is first added to the green liquor in a slaker tank, where it reacts with water to create calcium hydroxide. Liquor is fed from the slaker to agitated tanks, which enable the causticizing reaction to finish. The causticizing material is directed to the white liquor clarifier, which eliminates the calcium carbonate precipitate. To get rid of the last bits of sodium, this is put through the mud washer. The "reburned" lime is then added back to the slaker after being dried and calcined in the lime kiln using the mud. The recovery furnace smelt is dissolved in the SDT using the mud washer filtrate. The pulp digesters receive recycled white liquor from the clarifier (Bajpai 2008, 2015, 2018a; Tran 2007; Vakkilainen 2000; Venkatesh 1992; Adams 1992; Arpalahti et al. 2000; Reeve 2002).

2.4   Pulp Washing and Screening

Following production, pulp is cleaned to remove impurities and recycled through the pulp washing procedure any leftover cooking liquor. Washing pulp has a number of advantages. Screening, defibering, and deknotting are a few steps in pulp processing that eliminate impurities in the pulp. With the help of pulp washers, leftover spent pulping liquor is removed from the pulp. As the excess pulping liquor increases the requirement of bleach chemicals, efficient washing is essential to maximising pulping liquor return to chemical recovery and minimising carryover of pulping liquor into the bleach plant. In particular, the dissolved organic substances bind to the bleaching agents increasing their consumption.

To make the pulp more suitable for paper or board products, screening is used for removing oversized particles from the fine papermaking fibers. Knots are the largest particles in pulp. Knots are uncooked wood fragments, to put it simply. Prior to washing and fine screening, the knots are eliminated. These are broken in refiners in low yield pulps and are eliminated using specialised screens. Shives, are smaller fibre bundles. These do not get separated by mechanical or chemical action and are primarily removed through fine screening. Another type of oversized wood particles removed during screening is called chop. It is more problematic when digesting hardwood, as it usually arises from vessels and cells. The chopped particles are shorter and harder than shives. Debris is the term used for shives, wood chips and other materials which in some way adversely affect the paper properties (Ljokkoi 2000; Krotscheck 2006).

2.5   Bleaching

Without bleaching, mechanical pulp is used to create printing papers, primarily for newsprint, where low brightness is tolerable. But, the pulp needs to be bleached for the majority of printing, copying, and packaging grades. The majority of the lignin from the raw pulp is kept in mechanical pulps. It is bleached using hydrosulfites and hydrogen peroxides. The purpose of bleaching in the case of chemical pulps is to get rid of the tiny amount of lignin that is still present after cooking. Bleaching makes the pulp brighter so that it can be used to make tissue and printing grades of paper (Bajpai 2012).

Bleaching is done to remove lignin from wood for increasing the pulp brightness. Delignification and extraction stages for dissolved material alternate during the multistage process of bleaching. Delignification with oxygen or hydrogen peroxide can be used to strengthen the extraction process. Chemical Kraft bleaching, first used in the early twentieth century, has since been improved into a series of chemical reactions that happen gradually. The bleaching process has changed from a single stage using hypochlorite to a multistage process using chlorine, chlorine dioxide, oxygen, hydrogen peroxide, and ozone. Modern mills frequently use oxygen during the initial

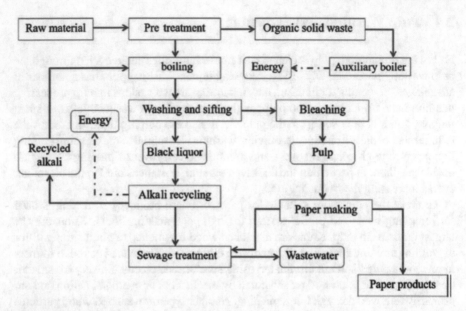

Fig. 2.1 Process flow diagram for pulp and paper production. Wang et al. (2016). Reproduced with permission

bleaching process. The current fashion is to bleach using totally chlorine-free (TCF) and elementally chlorine-free (ECF) methods. It is not advised to use elemental chlorine for bleaching. Only the ECF approach is suitable. The TCF method is preferred in terms of ecology.

The organic matter removed from the pulp during the bleaching with chlorine or chlorine chemicals and the material removed during the ensuing alkaline stages are chlorinated. Some of these organochlorine compounds are toxic and mutagenic (Bajpai 2012; Smook 1992; Reeve 1996a,b; McDonough 1992, 1995; Fredette 1996; Farr et al. 1992).

Figure 2.1 shows the process flow diagram for pulp and paper production (Wang et al. 2016).

2.6 Stock Preparation and Papermaking Process

Before pulp can be used to make paper, it must go through a process known as stock preparation. To transform raw stock into finished stock for the paper machine, stock preparation is done (Holik 2006). The preparation of the pulp for the paper machine involves mixing various pulps, diluting them, and adding chemicals. Chemical and mechanical pulps, recovered paper, and their mixtures are used as raw stocks. The characteristics of the paper produced are largely determined by the quality of the final stock. In the case of integrated mills, raw stock is offered in the form of loose material, bales or suspensions. Several procedures are involved in stock preparation, such as cleaning, fibre disintegration, fiber modification, storage and mixing.

Depending on the quality of the required furnishing and the raw stock used, these systems diverge significantly. When pulp is directly pumped in the pulp mill, the slushing stage and deflaking stage are skipped. The processes used are dispersion, beating/refining, metering, and blending of fibre and additives. Dry pulp is dissolved in water using a pulper for making a slurry. Refining is one of the most crucial processes in the preparation of fibres for papermaking. The processing of stock in a batch mode in the Hollander beater or one of its modified version is referred to as beating. When pulp is continuously fed to one or more refiners in series or in parallel, the process is called refining.

For a specific paper grade, refining creates different fibre properties in various ways. It increases the fiber's ability to bind while decreasing the drainage resistance without excessively damaging the fibers to reduce their individual strength. The finishing process is therefore based on the desired properties of the finished paper. Different fiber types respond in different ways due to their different morphological properties. In the finishing process, the type of fiber should be considered. Fibers are randomly and repeatedly subjected to tensile, compressive, shear and bending forces during the refining process (Bajpai 2005, 2018b; Smook 1992; Paulapuro 2000; Lumiainen 2000; Baker 2000, 2005; Stevens 1992).

Pulp are also treated with chemical additives. Resins are used for improving the wet strength of the paper. Fillers are used for improving the optical properties; dyes and pigments are used to affect the sheet color and sizing chemicals are used for improving the printing properties. Sizing chemicals control the liquid penetration (Neimo 2000; Davison 1992; Bajpai 2004; Krogerus 2007; Hodgson 1997; Roberts 1996, 1997).

After the stock preparation, the slurry is formed into the desired paper grade at the wet end of the paper machine. At this stage, the pulp is fed into the paper machine headbox. The slurry contains about 0.5% pulp fiber and remaining is water. The slurry outlet is the headbox opening. The fiber mixture is cast onto the moving wire mesh of a Fourdrinier machine or onto the rotating cylinder of a cylinder machine (Smook 1992).

The paper gets thicker, when more slurry is discharged from the headbox. Water is expelled through the screen as it moves along the path of the machine. To enhance sheet formation, the fibres entwine and align in the direction that the wire travels. After the formation of web the additional water is removed. A vacuum box under the wire helps with this drainage. In a Fourdrinier machine, the water is removed from one side of the sheet. Because of this, the sheet properties differ from one side to the other. This increases with the increase of machine speed. Twin-wire and multiple fourdrinier machines have been developed in response to this. Manufacturers of these machines use different vertical or horizontal designs. The paper web continues along the second wire after completing a brief forming run and loses moisture in the process. The next station is a press and dryer section where additional water is removed. The web first enters the press section and then enters the dryer. After entering the press, the paper is compressed between two rotating rollers and squeezes more moisture. The degree of water removed from the wire and press section is highly dependent on machine design and operating speed. The moisture in the sheet is around 65% after leaving the press. The paper web continues through the steam dryer and looses moisture in each step. Many tons of water evaporate in this process. Paper may be sized or coated. In such cases the web goes through a second drying cycle (part of the conversion cycle) before entering the calender stacker (part of the converting operation). The moisture level should be approximately 4 to 6% as specified at the factory. The paper becomes brittle if it is too dry. About 90% of the cost of moisture removal from the sheet takes place during pressing and drying. The energy required for drying accounts for most of the cost. The paper is fed onto rolls and wound up to the desired roll diameter, at the end of the paper machine. The machine tender cuts the paper at that diameter and starts a new roll immediately, the excess paper drops into a continuous web. This completes the process for the types of paper used to make corrugated board. For papers used for other purposes, the refining and converting operations usually take place outside the paper machine. These include coating, calendering, or supercalendering and winding. Coating is the surface treatment of the paper with pigments or adhesives for improving print quality, opacity, smoothness, color etc.

Paper with smooth printing surface have great demand. Several types of paper have a coating. Coated papers come in three broad categories: glossy, dull, and matt. For books and certain other products, coated paper having dull shade can be used for reducing glare from light while maintaining the benefits of coated paper. After the coating process, the film should be dried and re-rolled. Calendering is an on-machine process that runs the paper through the steel rollers to smooth the surface of the paper before it is wound into rolls. In addition to imparting smoothness, calendering can reduce film variability and produce a denser film. It also affects the water absorption properties of the paper.

Table 2.3 shows unit processes in stock preparation.

Table 2.3 Process steps in stock preparation

Unit process
Slushing and deflaking To break down the fiber raw material into a suspension of individual fibers. Slushing should at least result in a pumpable suspension enabling coarse separation and deflaking if required In the case of recovered paper, ink particles and other nonpaper particles should be detached from the fibers
Screening To separate particles from the suspension which differ in size, shape and deformability from the fibers
Fractionation To separate fiber fractions from each other according to defined criteria such as size or deformability of the fibers

(continued)

Table 2.3 (continued)

Unit process
Centrifugal cleaning To separate particles from the suspension which differ in specific gravity, size and shape from the fibers
Refining To modify the morphology and surface characteristics of the fibers
Selective flotation To separate particles from the suspension which differ in surface properties (hydrophobicity). from the fibers
Nonselective flotation To separate fine and dissolved solids from water
Bleaching To impart yellowed or brown fibers with the required brightness and luminance
Washing To separate fine solid particles from suspension (solid/solid separation).
Dewatering To separate water and solids
Dispersing To reduce the size of dirt specks and stickies (visibility, floatability), to detach ink particles from fibers

Based on Holik (2006)

References

Adams TN (1992) Lime reburning. In: Kocurek MJ (ed) Pulp and paper manufacture, 3rd edn. Joint committee of TAPPI and CPPA, vol 5, Atlanta, p 590

Arpalahti O, Engdahl H, Jantti J, Kiiskila E, Liiri O, Pekkinen J, Puumalainen R, Sankala H, Vehmaan-Kreula J (2000) Chapter 14: white liquor preparation. In: Gullichsen J, Paulapuro H (eds) Papermaking science and technology book 6B, Fapet Oy, p 135

Arppe M (2001) Mechanical pulp: has it got a future or will it be discontinued? Int Papwirtsch 10:45–50

Bajpai P (2004) Emerging technologies in sizing. PIRA International, U.K., p 159

Bajpai P (2005) Technological developments in refining. PIRA International, U.K., p 140

Bajpai P (2008) Chemical recovery in pulp and paper making. PIRA International, U.K., p 166

Bajpai P (2012) Environmentally benign approaches for pulp bleaching, 2nd edn. Elsevier, BV, Amsterdam

Bajpai P (2013) Recycling and deinking of recovered paper. Elsevier Science, Amsterdam

Bajpai P (2015) Green chemistry and sustainability in pulp and paper industry. Springer International Publishing, Cham, p 258

Bajpai P (2018a) Biermann's handbook of pulp and paper: volume 1: raw material and pulp making. Elsevier, USA

Bajpai P (2018b) Biermann's handbook of pulp and paper volume 2: paper and board making. Elsevier, USA

Baker C (2000). Refining technology. In: Leatherhead BC (ed). PIRA International, UK, p 197

Baker CF (2005). Advances in the practicalities of refining. In: Scientific and technical advances in refining and mechanical pulping, 8th PIRA international refining conference, PIRA International, Barcelona, Spain (28 Feb–Mar 2005)

Davison RW (1992). Internal sizing. In: Hagemeyer RW, Manson DW (eds). Pulp and paper manufacture, vol 6. Joint Text Book Committee of TAPPI and CPPA, p 39

EPA (2001a) Pulp and paper combustion sources national emission standards for hazardous air pollutants: a plain english description. U.S. Environmental Protection Agency. EPA-456/ R-01–003. Sept 2001a. http://www.epa.gov/ttn/atw/pulp/chapters1-6pdf.zip

EPA (2001b) Pulping and bleaching system NESHAP for the pulp and paper industry: a plain english description. U.S. Environmental Protection Agency. EPA-456/R-01–002. Sept 2001b. http://www.epa.gov/ttn/atw/pulp/guidance.pdf

Farr JP, Smith WL, Steichen DS (1992) Bleaching agents (survey). In: Grayson M (ed) Kirk-Othmer encyclopedia of chemical technology, vol 4, 4th edn. Wiley, New York, p 271

Fredette MC (1996) Pulp bleaching: principles and practice. In: Dence CW, Reeve DW (eds) Bleaching chemicals: chlorine dioxide (Section 2, Chapter 2). Tappi Press, Atlanta, p 59

Gullichsen J (2000) Fibre line operations. In: Gullichsen J, Fogelholm C-J (eds) Chemical pulping—papermaking science and technology. Fapet Oy, Helsinki, Finland, Book 6A, p A19

Hodgson KT (1997) Overview of sizing. Tappi Sizing Short Course, Session 1, Nashville, TN (Apr 14–16, 1997)

Holik H (2006) Stock preparation. In: Sixta H (ed) Handbook of paper and board. WILEY-VCH Verlag GmbH & Co. KgaA, pp 150–206

Krogerus B (2007) Chapter 3: Papermaking additives. In: R Alen (ed) Papermaking chemistry: papermaking science and technology book 4, 2nd edn, pp 54–121. Finnish Paper Engineers' Association, Helsinki, Finland, p 255

Krotscheck AW (2006) In: Sixta H (ed) Handbook of pulp. WILEY-VCH Verlag GmbH & Co. KgaA, pp 512–605

Ljokkoi R (2000) Pulp screening applications. In: Gullichsen J, Fogelholm C-J (eds) Papermaking science and technology, vol 6A. Chemical Pulping. Fapet Oy, Helsinki, pp A603–A616

Lumiainen J (2000) Chapter 4: Refining of chemical pulp. In: Papermaking science and technology, papermaking part 1: stock preparation and wet end, vol 8, p 86. Fapet Oy, Helsinki, Finland

McDonough T (1992) Bleaching agents (pulp and paper). In: Grayson M (ed) Kirk-Othmer encyclopedia of chemical technology, vol 4. Wiley, New York, p 301

McDonough TJ (1995) Recent advances in bleached chemical pulp manufacturing technology. Part 1: extended delignification, oxygen delignification, enzyme applications, and ECF and TCF bleaching. Tappi J 78(3):55–62

Neimo L (2000) Internal sizing of paper. In: Neimo L (ed) Papermaking chemistry. Tappi Press, Fapet Oy, Helsinki, Finland, p 150

Paulapuro H (2000) Stock and water systems of the paper machine. In: Gullichsen J, Fogelholm C-J (eds) Papermaking part 1: stock preparation and wet end—papermaking science and technology, Fapet Oy, Helsinki, Finland, Book 8, p 125

Reeve DW (1996a) Introduction to the principles and practice of pulp bleaching. In: Dence CW, Reeve DW (eds) Pulp bleaching: principles and practice. Tappi Press, Atlanta, p 1, Section 1, Chapter 1

Reeve DW (1996b). Pulp bleaching: principles and practice. In: Dence CW, Reeve DW (eds) Chlorine dioxide in bleaching stages. Tappi Press, Atlanta, Section 4, Chapter 8, p 379

Reeve DW (2002) The Kraft recovery cycle. Tappi Press, Tappi Kraft Recovery Operations Short Course

Ressel JB (2006) In: Sixta H (ed). Handbook of pulp. WILEY-VCH Verlag GmbH & Co. KgaA, pp 69–105

Roberts JC (1996) Neutral and alkaline sizing. In: Roberts JC (ed) Paper chemistry, 2nd edn. Chapman and Hall, London, UK, p 140

Roberts JC (1997) A review of advances in internal sizing of paper. In: Baker CF (ed) The fundamentals of paper making materials, transactions, 11th fundamental research symposium (Cambridge), vol 1, p 209

Sixta H (2006) In: Sixta H (ed). Handbook of pulp. WILEY-VCH Verlag GmbH & Co. KgaA, p 2–19

Smook GA (1992) Handbook for pulp and paper technologists. Joint textbook committee of the paper industry of the United States and Canada, p 425

Stevens WV (1992) Refining. In: Kocurek MJ (ed) Pulp and paper manufacture, vol 6, 3rd edn. Joint committee of TAPPI and CPPA, Atlanta, p 187

Tran H (2007). Advances in the Kraft chemical recovery process. Source 3rd ICEP international colloquium on eucalyptus pulp, 4–7 Mar, Belo Horizonte, Brazil, p 7

Vakkilainen EK (2000) Chapter 1: chemical recovery. In: Gullichsen J, Paulapuro H (eds) Papermaking science and technology book 6B, Fapet Oy, p 7

Venkatesh V (1992) Chapter 8: lime reburning. In: Greenand RP, Hough G (eds) Chemical recovery in the alkaline pulping process 6B. Tappi Press, p 153

Wang Y, Yang X, Sun M, Ma L, Li X, Shi L (2016) Estimating carbon emissions from the pulp and paper industry: a case study. Appl Energ 184:779–789

Chapter 3
Environmental Issues of the Pulp and Paper Industry

Abstract The discharge of wastewater, environmentally friendly waste management, energy conservation and recovery, and the local odor from kraft pulp mills are anticipated to remain top environmental action priorities in the future by the pulp and paper industry. The nature of the effects caused by issues arising from the paper and pulp industry is used to classify them. This includes deforestation; air emissions; water pollution and sludge and solid waste.

Keywords Pulp and paper industry · Deforestation · Air emissions · Water pollution · Sludge and solid waste

The pulp and paper industry is resource and capital-intensive industry that contributes to a number of environmental issues, such as acidification, nutrification, acidification of the atmosphere, global warming, human and eco-toxicity and solid waste (Bajpai and Bajpai 1996; Patrick 1997). The pulping and bleaching processes have the biggest effects on the environment because they release pollutants into the air, wastewater, and solid waste. Since this part of the manufacturing process has historically been linked to the formation of organochlorine compounds, much research has concentrated on the bleaching techniques used. These pollutants can contaminate the food chains by bioaccumulation and are non-biodegradable, toxic, and persistent. The dioxins are thought to cause cancer and are notorious for their extreme toxicity (Söderholm et al. 2019).

Another important factor affecting the plant's ability to close the process loop and achieve zero-wastewater operation is bleaching technology (Gleadow et al. 1997). Pulp and paper mills are making efforts to, close water circuits. So, a further reduction in emissions are anticipated (toward zero wastewater mills). But, at present there are no kraft mills in operation that fully recover the effluent from the bleach plant. A handful of CTMP mills, a Na-based sulfite pulp mill, and a test liner and corrugated manufacturer using recycled fibers have achieved zero wastewater discharge (Bajpai 2001). Recently, pulp production with minimal impact has been discussed.

This minimal impact factory represents broader issues and challenging concepts such as minimizing the emissions and resource consumption, minimizing cross-media effects, and considering work environments and the economic aspects (Hanninen 1996; Elo 1995).

© The Author(s), under exclusive license to Springer Nature Switzerland AG 2023
P. Bajpai, *Environmentally Benign Pulping*, SpringerBriefs in Green Chemistry for Sustainability, https://doi.org/10.1007/978-3-031-23693-8_3

Historically, atmospheric sulfur emissions from pulp mills have been very high, but improvements in process technology have reduced atmospheric sulfur emissions significantly, especially in recent years.

The recycling of used paper fibres has advanced to a fairly advanced level in the majority of the countries, and for some paper grades, further growth is possible. Rejects and sludge from pulp and paper manufacturing processes can be used for recovering energy, preventing the need for waste disposal. However, there is still a significant opportunity to use effective on-site methods more frequently in this regard. Although no external energy is required for chemical pulping, the overall demand for process energy remains very high.

Mechanical pulping is the most energy intensive process due to refiner power requirements. Papermaking and the processing of recovered paper both require a lot of energy. This is because pressing and drying must be used to increase the solids content of the dilute suspension of fibers and possibly fillers to approximately 95% solids. This is the typical dry solids content of finished paper.

Prior to the 1970s, the pulp and paper industry had a significant impact on water intake with wastewater discharge. The effects, which included fish kills and oxygen depletion, were occasionally quite dramatic. From the late 1970s until now, the main focus has been on the role of organochlorine compounds produced in the bleach plants. Over the past decade, the use of elemental chlorine in bleaching has declined dramatically. This is due to public concerns over the potential environmental hazards posed by using chlorine in the bleaching process (Bajpai 2001).

Stricter environmental regulations have resulted from growing awareness of the effects bleach effluent has on the environment. The bleach plant's effluent was able to be recycled back into the chemical recovery system of the mill, allowing the mill system to be shut down and the chloride content of the effluents to be significantly reduced. By using in-process measures like these, it has been possible to significantly reduce chlorinated as well as non-chlorinated organic compounds in the pulp mill effluents (McDonough 1995). Examples include increased delignification prior to the bleach plant through modified or added pulping processes, additional oxygen stages, spill recovery techniques, effective washing techniques, and condensate stripping and recycling. Unchlorinated toxic organic compounds and AOX emissions have decreased, which is another contributing factor. The pulp and paper industry is currently trending in the direction of more bleach plants being closed, either through pulp bleaching using ECF or TCF (Pryke 2003; Bajpai 2012). By putting in place advanced wastewater treatment systems, integrated in to production, paper mills can reuse treated process waters more frequently. The pulp and paper industry expects that the discharge of wastewater, environmentally friendly waste management, energy conservation and recovery, and the local odour from kraft pulp mills will continue to be top priorities for environmental action in the future. Problems arising from the paper and pulp industry are characterized based on the nature of their impact. This includes the followings issues.

3.1 Deforestation

The number of trees is declining as a result of the increase in daily paper consumption. According to calculations, paper use has increased 400% globally over the past 40 years. The percentage of CO_2 in the atmosphere will rise as a result of the decline in tree cover, which will lead to global warming and a change in the climate. Disasters like drought and flooding are also caused by the ratio of agricultural land decreasing. Additionally, as the number of trees declines, the amount of oxygen in the atmosphere also decreases.

3.2 Air Emissions

Particulates, hydrogen sulphide, sulphate and nitrate oxides, as well as particulates, are the main components of chemical pulp mill air emissions. Chloroform, furans, dioxins, and other organochlorine and other volatile organic compounds (VOCs) are examples of micropollutants. Similar to liquid effluent discharges, the type of process technology used and specific mill practices have a significant impact on the levels of emissions. The type and quality of fuel is another crucial factor. The majority of harmful gas and particulate emissions from older mills have been eliminated by mitigation technology. Depending on regional elements like local laws, business and mill policies, and proximity to populated areas, this technology may or may not be used.

Discussions about the pulp and paper industry's role in global warming have been going on for a while. Some contend that the greenhouse gas emissions from the production, transportation, and disposal of pulp and paper products are more than offset by the carbon dioxide that plantation forestry absorbs. The International Institute for Environment and Development's research disputes this assertion.

The study found that recycling paper provides a net increase of about 450 million carbon dioxide equivalent units annually. Although the benefits of process changes in plants regarding water resource utilization and the associated impacts on water systems are well documented, the impact of atmospheric emissions (mainly from recovery boiler) due to closed-loop operation is significant. The impact of the change is not well documented.

The possibility of incomplete combustion byproducts and other potentially dangerous substances, such as chemicals like dioxins, coming from ECF mills is obviously a cause for concern. Fly ash produced by burning kraft mill sludge has been found to contain PCBs, dioxins, and furans, raising concerns that significant amounts may be emitted into the atmosphere. According to a study conducted in British Columbia, Canada, flue gases from recovery boilers having higher chloride loads may not be the primary source of dioxin and furan emissions into the atmosphere, but other recovery boilers were found to contain these persistent organic pollutants (Environment Canada 1998; Luthe et al. 1997; Kopponen 1994).

Given the potential threat to the health and safety of factory workers, some hazardous air pollutants (HAP), trace air pollutants and overall reduced sulfur compounds are subject to individual regulations and controls. A "light stripper" was installed in one plant to clean the less contaminated condensate in the evaporator stage.

Malodorous gases can be gathered by the inadequate gas system and burned in the recovery boiler. This reduces offensive discharges and sulphur emissions. Other significant air pollutants from recovery boilers that warrant concern include hydrogen chloride and methanol. These pollutants are released in greater amounts by older direct contact evaporator recovery boilers, which also produce significant sulphur emissions. Installation should ideally be a part of the process of converting mills to closed loop operation (Södra Cell 1996; Andrews et al. 1996).

Methanol is produced during the process and released from oxygen delignification and white liquor oxidation systems, along with a variety of other dangerous air pollutants and VOCs. Methanol emissions from oxygen delignification systems are likely to be significantly benefited by lowering the content of methanol of the last post-oxygen washer shower water. Pulp mill air emissions have been linked and VOCs (NCASI 1994). There is no clear link between decreased methanol emissions and decreased emissions of other pollutants found in real operating milto a variety of HAPsl environments, such as methyl mercaptan and chlorobenzene. Similar to methanol, phenols do not seem to be reduced proportionally.

TCF mills do not produce any chlorinated compounds during bleaching operations, whereas ECF mills produce chlorinated compounds. Chloroform, methyl dichloroacetate, 2,5-dichlorothiophane, and other VOCs have been identified in plant off gases using 100% chlorine dioxide substitution.

These substances were not discovered in a TCF mill, but it has been discovered that they get volatilized from the treatment ponds of these mills. Chlorine dioxide use has significantly reduced the concentrations of chlorinated compounds compared to levels produced by mills using elemental chlorine, but it has not completely eliminated them. Due to their release into the local environment and potential for long distance transportation, these chemicals have an impact on the environment (Juuti et al. 1996; Calamari et al. 1994).

Chlorine dioxide is an air pollutant is of major concern, particularly in relation to potential leaks and fugitive emissions within factories (Simons 1994). The main advantage that TCF mills do not produce HAPs in the bleach plant has been acknowledged by the US EPA (1998). The technology used in the modern mills generally has positive environmental effects on air emissions, and non-chlorine chemical bleach sequences have additional advantages.

3.3   Water Pollution

Bleaching plant effluents are the dirtiest wastewaters produced by the pulp and paper industry. The effluents from bleached pulp mills contain about 300 different compounds that have been identified. About 200 of them are dioxins, chlorinated phenolics, chlorinated resin acids, and other chlorinated organic compounds. Dioxins and chlorinated phenols are two toxic, nonbiodegradable pollutants that have a propensity to bioaccumulate and contaminate food chains. Dioxins are notorious for being extremely toxic (Bajpai 2001).

The effluent from bleach kraft mills contains chlorinated and non-chlorinated lignin products as well as wood extractives, giving it a dark colour. Using coloured effluent can have a number of negative effects on the receiving water body (Bajpai 2001).

Bleached wastewater from kraft mill affects the biological quality of the water received. Loss of benthic invertebrates, higher rates of fish diseases, and mutagenic effects on aquatic animals are some of the consequences of discharge of bleaching effluents into surface waters. Bleached kraft effluent and sulfite effluent damage the function of the liver, enzymatic systems, and metabolic cycles of the exposed fishes. The low molecular weight components of bleach effluents contain problematic compounds which enters the cell membranes and bioaccumulate. These organochlorine compounds have a significant impact on the biology of aquatic ecosystems (Sundelin 1988).

3.4 Sludge and Solid Waste

Pulp production generates large amounts of solid waste, including wood waste (mainly bark), sodium salts from recovery boilers, waste from pulp sorting, and dregs and grit from the causticizing plants. Additionally, when wood waste and sludge are burned, ashes are produced. Bark, sawdust, and other wood fragments are examples of the most significant pulp mill residue: wood waste. In a boiler, wood waste is burned to create energy for the mill. The same boiler is used to burn both wood waste and pulp screening rejects. Landfills are used to dispose of the causticizing plant's dregs and grit (Gavrilescu 2004). The sludge produced during wastewater treatment is the second residue. Depending on the paper grade, sludge volume varies significantly.

Sludge production from the manufacture of paper and the production of pulp together totals 20–25 kg/t. If recovered paper is used to make the paper, there will be an increase in sludge of up to 150 kg/t of paper. Sludge is burned after dewatering (Gavrilescu 2005).

Although landfills are typically where solid waste is disposed of, incineration is becoming more common. Other methods include using the waste to improve soil. However, as with all disposal methods, there is some worry about potential contamination with heavy metals and dioxin. Problems with disposing of solid waste are greatly reduced in a closed loop mill. But, purge points will be necessary for controlling the nonprocess elements in order to avoid chemistry hiccups in the bleaching and recovery processes and reducing the corrosion of mill machinery.

   As sludge and solid waste continue to be generated, the quality of this waste becomes a major concern. This is especially true given how frequently land spreading is advocated as a method of getting rid of these wastes. Sludge that has not been contaminated may prove to be a useful resource. The reuse of otherwise unrecyclable wastes might be facilitated by composting of properly treated sludge. Another desirable objective might be to use the waste products from pulping and bleaching as raw materials for other processes. Various chemicals, both of natural origin and those created entirely as a result of the pulping and bleaching processes, can be found in the sludge from bleach kraft pulp mills. It is well known that the industry's adopted process changes affected the sludge's quality (Paleologou et al. 1997).

   Sulfate compounds are a by-product of chlorine dioxide production and are often used as make-up in pulping and so closing the ECF sludge loop increases the disposal of sulphur chemicals from the chlorine dioxide generator. Chemicals containing sulphur are more concentrated in process circuits due to increased production of chlorine dioxide for ECF and increased filtrate recycling. It is necessary to dispose of extra sulphates because increased sulphur raises concerns for non-process element control in closed loop designs. In many current mills, these eventually make it into the effluent treatment process. Assertions that increased effluent recycling will lead to an eventual doubling of lime muds, dregs, precipitator ash and other purge streams must be viewed with some concern (Ryynänen and Nelson 1996).

   Currently, grits, dregs, and ash make up approximately 3% of the material that has been dissolved as a result of pulping and bleaching operations. Closed loop operations may increase that percentage to 6%, but this should be compared to the total elimination of liquid effluent discharge, dissolved waste fibre, and spent liquors that are discharged into water or on land.

References

Andrews RS, Paisley MA, Smith M, Young RJ (1996) Formation and release of methanol and hydrogen chloride from DCE recovery boiler systems. Tappi J 79(8):55–160

Bajpai P (2001) Microbial degradation of pollutants in pulp mill effluents in advances in applied microbiology. In: Neidleman S, Laskin A (eds), Academic Press—New York), vol 48, pp 79–134

Bajpai P, Bajpai PK (1996) Organochlorine compounds in bleach plant effluents—genesis and control. PIRA Internationa, Leatherhead, Surrey, UK

Bajpai P (2012) Environmentally benign approaches for pulp bleaching, 2nd edn. Elsevier, BV, Amsterdam

Calamari D, Tremolada P, Guardo AD (1994) Chlorinated hydrocarbons in pine needles in Europe: fingerprint for the past and recent use. Environ Sci Technol 28:429–434

Cell S (1996) Annual environmental report. Södra Cell, Sweden

Elo A (1995) Minimum impact mill, third global conference on paper and the environment, London, UK, pp 105–110, 26–28 Mar 1995

Environment Canada (1998) Dioxins and furans and hexachlorobenzene inventory. Environment Canada and the federal/provincial task force on dioxins and furans for the federal-provincial advisory committee for the Canadian environmental protection Act Draft 4. April 1998

Gavrilescu D (2004) Solid waste generation in kraft pulp mills. Environ Eng Manag J 3:399–404

Gavrilescu D (2005) Best practices in Kraft pulping—benefits and costs. Environ Eng Manag J 4:29–46

Gleadow P, Hastings C, Richardsdon B, Towers M, Uloth V (1997) Towards closed-cycle kraft: ECF versus TCF case studies. Pulp Pap Can 98:4

Hanninen E (1996) Minimum impact mill. Paper Asia 12(1):24–29

Juuti S, Vartiainen T, Joutsenoja P, Ruuskanen J (1996) Volatile organochlorine compounds formed in the bleaching of pulp. Chemosphere 33(3):437–448

Kopponen P, Välttilä O, Talka E, Törrönen R, Tarhanen J, Ruuskanen J, Kärenlampi S (1994) Chemical and biological 2,3,7,8-tetrachlorodibenzo-p-dioxin equivalents in fly ash from combustion of bleached Kraft pulp mill sludge. Environ Toxicol Chem 13:143–148

Luthe CE, Uloth V, Karidio I, Wearing J (1997) Are salt-laden recovery boilers a significant source of dioxins?". Tappi J 80(2):165–169

McDonough TJ (1995) Recent advances in bleached chemical pulp manufacturing technology. Part 1: extended delignification, oxygen delignification, enzyme applications, and ECF and TCF bleaching. Tappi J 78(3):55–62

NCASI (National Council of the Pulp and Paper Industry on Air and Stream Improvement) (1994) Technical Bulletin #684." NCASI New York

Paleologou M, Thibault A, Wong PY (1997) Enhancement of the current efficiency for sodium hydroxide production from sodium sulphate in a two-compartment bipolar membrane electrodialysis system. Sep Purif Technol 11:159-171

Patrick KL (1997) Advances in bleaching technology. Miller Freeman Books, CA, p 20

Pryke D (2003) ECF is on a roll: it dominates world bleached pulp production. Pulp Pap Int 45(8):27–29

Ryynänen H, Nelson P (1996) Environmental life cycle assessment of some new methods for producing bleached pulps from Australian eucalypt woods. Appita 49(3):167–172

Simons HA (1994) Technical background information document on pulp and paper mill air emissions. H.A. Simons Ltd. P.5517 A

Söderholm P, Bergquist AK, Söderholm K (2019) Environmental regulation in the pulp and paper industry: impacts and challenges. Curr Forestry Rep 5:185–198

Sundelin B (1988) Effects of sulphate pulp mill effluents on soft bottom organisms—a microcosm study. Water Sci Technol 20(2):175–177

US EPA (1998) Maximum achievable control technology (MACT) I & II. Title 40, Chapter 1, CFR, Part 63

Chapter 4
Concerns of the Conventional Pulping Methods

Abstract Pulp and Paper industry poses a threat to the environment and has a significant impact on human health, the quality of air and water, and natural ecosystems. Economic conditions and environmental pressures have had a significant impact on the pulp and paper industry, which is under intense pressure to reduce pollution emissions. Concerns about the environment are escalating, and as a result, pressure is mounting to implement cutting-edge green technologies. During the production of pulp and paper, the processes of wood pulping and pulp bleaching have the greatest negative effects on the environment. Chemical pulping is the primary industry contributor to air pollution because it operates at higher temperatures. The concerns of the conventional pulping methods are discussed in this chapter.

Keywords Pulp and paper industry · Chemical pulping · Air pollution · NOx emissions · TRS emissions · TRS odours · Methyl mercaptan · Hydrogen sulphide · Dimethyl disulphide · Dimethyl sulphide

The pulp and paper industries have been identified as one of the key industries that threatens the environment and has a significant impact on human health, air and water quality, and natural ecosystems. The pulp and paper industry has been severely impacted by economic conditions and environmental pressures, and it is under intense pressure to improve performance in relation to the release of pollutants. Pressure to adopt new, environmentally friendly technologies is growing due to growing environmental concerns. The wood pulping and pulp bleaching processes have the biggest negative effects on the environment during the production of pulp and paper. The industry's primary source of air emissions is chemical pulping, primarily because it operates at higher temperatures (EPA 2001a; European Commission 2001).

The storage of chips, cooking, washing of pulp, preparation of bleaching chemicals, recovery of chemicals, evaporation plant, bark furnace, recovery and auxiliary boilers, preparation of white liquor, lime kiln, storage tanks, and, in the case of market pulp, drying of pulp are all sources of air emissions during chemical pulping (Gullichsen 2000; Bajpai 2008). Recovery boilers are a major source of air emissions in pulp mills (Adam et al., 1997). The majority of these emissions are sulphur

Table 4.1 The main sources of TRS emissions

Digester Methyl mercaptan, methanol
Black liquor storage tank Hydrogen sulphide, methyl mercaptan, dimethyl sulphide
Evaporator Hydrogen sulphide, dimethyl disulphide, dimethyl sulphide, methanol
Recovery boiler Hydrogen sulphide, methyl mercaptan, dimethyl sulphide
Smelt dissolving tank Hydrogen sulphide, methyl mercaptan
Lime kiln Methyl mercaptan, sulphur dioxide

dioxide, but there are also nitrogen oxides, smelly compounds, and particulate emissions. NOx emissions are influenced by the supporting fuel rate used in the recovery boiler and the nitrogen content of the black liquor.

In the kraft process, wood chips are cooked under pressure with sodium sulfide and sodium hydroxide in a digester for separating the lignin and cellulose from the chips (Bajpai 2008). After filtering, washing, bleaching, pressing, and drying, the pulp is transformed into paper. The sodium sulfide in the white liquor serves as a source of sulphur for the TRS compounds. Hydrogen sulfide is typically the major TRS compound released from this process, but many other TRS compounds are also produced during pulping as the sodium sulfide reacts with the lignin and the process gases. Methyl mercaptan is one of these chemicals. In order to help detect gas leaks, natural gas companies intentionally add a strong-smelling substance to odourless natural gas (Smet et al. 1998; Das and Jain 2004).

These four substances are released from various locations throughout a mill and have low thresholds for odour. The direct contact evaporator and digester/blow tank systems are the main sources. From the digester, the majority of TRS compounds are released, other parts of the process can also be sources of TRS emissions. Sometimes unusually potent but brief TRS odours can be detected near these plants as a result of temporary process disruptions or even changes in weather (Bajpai 2014). The primary sources of TRS emissions are digester blow and relief gases, recovery furnaces with direct-contact evaporators, multiple-effect evaporator vents and condensates, smelt dissolving tanks, and slacker vents (Table 4.1) (Anderson 1970; Andersson et al. 1973; Bordado and Gomes, 1997; 2003; EPA 2001b).

Millions of people around the world are affected by atmospheric pollution, especially those living in large industrialised cities because of the offensive smell, fumes, dust, and corrosive gases that are bad for people's health, crops, and property (Bajpai 2014). The receptiveness of people to the effects of air pollution varies greatly, according to reports. The elderly, children, and the ill are relatively much more sensitive, while healthy adults can tolerate relatively higher concentrations of detrimental substances without experiencing any harm. Since hydrogen sulfide is both the most toxic and the largest proportional component of TRS, it has received a lot of attention in research. Hydrogen sulfide, methyl mercaptan, dimethyl sulfide, and dimethyl disulfide have lower detectable limits of 8 to 20, 2.4, 1.2, 15.5 respectively.

The smell of hydrogen sulfide is often described as "rotten eggs." Low concentrations of—0.001 to 0.13 ppm—can be detected. Through the lungs, hydrogen sulfide is easily absorbed. When hydrogen sulfide interacts with cytochrome oxidase, it prevents cells from absorbing oxygen (Hessel et al. 1997). When exposure levels are greater than 100 ppm, inflammation and irritation in eye are seen within 2 to 15 min. Irritation in eyes has been noted at lower concentrations but it is unclear whether the effects of hydrogen sulfide alone or or together with exposure to other gases are answerable (10 ppm). Due to the possibility of olfactory nerve fatigue after exposure to high levels, the smell is a poor indicator of the presence of hydrogen sulfide (100 pm between 2 to 15 min or continued exposure). The throat, nose, and lungs may become irritated at concentrations of or beyond 100 ppm. Olfactory pulmonary oedema has been linked to concentrations over 250 ppm. An exposure level over 500 ppm may cause unconsciousness. At higher concentrations (500 to 1000 ppm), *deficiency of oxygen reaching the tissues,* respiratory and cardiovascular effects can occur.

People experiencing acute exposure to hydrogen sulfide have described a number of long-term, persistent health effects. The most common symptoms are neurological ones like headaches, memory loss, difficulty focusing, respiratory ones like wheezing and shortness of breath, and eye diseases.

Following up on people who had been exposed to acute levels of hydrogen sulfide at work revealed reports of damage to brain structures. Studies on the workplace have revealed evidence of harmful health effects, such as mood disorders and bronchial hyper responsiveness. Communities exposed to high short-term or low long-term levels of industrial emissions are shown to have long-term negative health effects. Local population exposed to ambient TRS and/or hydrogen sulfide showed some signs of respiratory and central nervous system effects. Eye and nasal irritation, shortness of breath, cough and nausea are among the symptoms. Due to the numerous flaws in these studies, it is challenging to assess the extent of health risk for community members exposed to TRS and/or hydrogen sulfide. It is not easy to distinguish between acute and chronic exposure-related effects because of the potential for chronic exposure in communities. The central nervous system is affected by methyl mercaptan. It has been discovered that at high concentrations, it can cause convulsions and narcosis in addition to paralysing the respiratory centre. In smaller doses, it results in pulmonary congestion (Hessel et al. 1997). Studying the case of a man who became overexposed while emptying methyl mercaptan gas cylinders, as reported by Shults et al. (1970), makes it simple to comprehend the toxin's effects. At the job site, he was discovered unconscious and taken to the hospital. He experienced methemoglobinemia and acute hemolytic anaemia, and died 28 days after the accident while still unconscious (AMEC 2004; ATSDR, 2006, 2011; BC MOE 2009; Campagna et al. 2004; CCOHS 2012; Hessel et al. 1997; Inserra et al. 2004; Ontario MOE 2007; Shults et al. 1970; Slaughter et al. 2003; Smet et al. 1998; US EPA 2003).

From NSSC pulping process, the emissions are generally much lower than emissions from Kraft process. Since there is no sodium sulfide in the cooking liquor, both methyl mercaptan and dimethyl sulfide are not present in gaseous emissions, very little reduced sulfur is released. From the sodium carbonate process, the emissions of sulfur are due to sulfur in the fuel oil and process water streams. Emissions of sulfur dioxide and nitrogen oxides are similar to those of kraft mill (Dallons 1979).

Typical emissions in a sulfite process are sulfur dioxide with certain nitrogen oxides. Sulfur dioxide is also released during preparation of sulfite liquor and recovery. There are almost no sulfur dioxide emissions from continuous digesters. However, batch digesters can release large amounts of sulfur dioxide depending upon how the digester is emptied. Emissions from sulfite digesters and blow-pit depend upon the type of system used. These areas can be major sources of sulfur dioxide emissions. Sulfur dioxide is also emitted from the pulp washers and multiple-effect evaporators.

Emissions to air from mechanical pulping processes are primarily related to the emissions of VOCs. Sources of VOCs are air emissions from raw material washing chests and other chests, and from sparkling washer where steam released contaminated with wood volatiles are condensed. VOC level depends upon the quality and freshness of the raw materials and the methods used.

These constraints associated with usual pulping methods can be overcome with the use of environmentally benign pulping technologies.

References

Adams TN, Frederik WJ, Grace TM (1997) Kraft recovery boilers. TAPPI, Atlanta (USA 99)

AMEC, University of Calgary (2004) Assessment report on reduced sulphur compounds for developing ambient air quality objectives. http://environment.gov.ab.ca/info/library/6664.pdf

Anderson K (1970) Formation of organic sulfur compounds during Kraft pulping II. Influence of some cooking variables on the formation of organic sulfur compounds during kraft pulping of Pine. Svensk Paperstid 73(1):1

Andersson B, Lovblod R, Grennbelt P (1973) Diffuse emissions of odourous sulphur compounds from kraft pulp mills, 1 VLB145. Swedish water and air pollution research laboratory, Gotenborg

ATSDR (Agency for Toxic Substances and Drug Registry) (2006) Toxicological profile for hydrogen sulfide. http://www.atsdr.cdc.gov/toxprofiles/tp114.pdf

ATSDR (Agency for Toxic Substances and Drug Registry) (2011) Medical management guidelines for hydrogen sulfide. http://www.atsdr.cdc.gov/mmg/mmg.asp?id=385&tid=67

Bajpai P (2008) Chemical recovery in pulp and paper making. PIRA International, U.K, p 166

Bajpai P (2014) Biological odour treatment: Springer 2014:20–32 https://link.springer.com/content/pdf/10.1007/978-3-319-07539-6.pdf

BC Ministry of Environment (2009) Air quality objectives and standards. http://www.bcairquality.ca/reports/pdfs/aqotable.pdf

Bordado JCM, Gomes JFP (1997) Pollutant atmospheric emissions from Portuguese Kraft pulp mills. Sci Total Environ 208(1–2):139–143

Bordado JCM, Gomes JFP (2003) Emission and odour control in kraft pulp mills. J Clean Prod 11:797–801

Campagna D, Kathman SJ, Pierson R, Inserra SG, Phifer BL, Middleton DC (2004 Mar) Ambient hydrogen sulfide, total reduced sulfur, and hospital visits for respiratory diseases in northeast Nebraska, 1998–2000. J Expo Anal Environ Epidemiol 14(2):180–187

Canadian Centre for Occupational Health and Safety (2012) Hydrogen sulfide. http://www.ccohs.ca/products/databases/samples/cheminfo.html. Accessed 20 March 2013

Dallons V (1979) Multimedia assessment of pollution potentials of non-sulphur chemical pulping technology. Environmental protection agency, office of research and development, industrial environmental research laboratory, Cincinnati (EPA-600/2–79-026, January 1979)

Das TK, Jain AK (2004). Pollution prevention advances in pulp and paper processing. Environ Prog 20(2):87–92

EPA (2001a) Pulping and bleaching system NESHAP for the pulp and paper Industry: a plain English description. U.S. environmental protection agency. EPA-456/R-01–002. September 2001a. http://www.epa.gov/ttn/atw/pulp/guidance.pdf

EPA (2001b) Pulp and paper combustion sources national emission standards for hazardous. U.S. environmental protection agency, office of air and radiation, office of air quality planning and standards, Washington, DC

European Commission (2001) Integrated pollution prevention and control (IPPC). Reference document on best available techniques in the pulp and paper industry. Institute for Prospective Technological Studies, Seville

Gullichsen J (2000) Fibre line operations. In: Gullichsen J, Fogelholm C-J (eds) Chemical pulping—papermaking science and technology. Fapet Oy, Helsinki, p A19 (Book 6 A)

Hessel PA, Herbert FA, Melenka LS, Yoshida K, Nakaza M (1997) Lung health in relation to hydrogen sulfide exposure in oil and gas workers in Alberta, Canada. Am J Ind Med 31(5):554–557

Inserra SG, Phifer BL, Anger WK, Lewin M, Hilsdon R, White MC (2004) Neurobehavioral evaluation for a community with chronic exposure to hydrogen sulfide gas. Environ Res 95(1):53–61

Shults WT, Fountain EN, Lynch EC (1970) Irreversible coma and hemolytic anemia following inhalation. J Am Med Assoc 211:2153–2154

Slaughter JC, Lumley T, Sheppard L, Koenig JQ, Shapiro GG (2003) Effects of ambient air pollution on symptom severity and medication use in children with asthma. Annal Allergy Asthma Immunol 91(4):346–353

Smet E, Lens P, Van Langenhove H (1998) Treatment of waste gases contaminated with odorous sulfur compounds. Critical Rev Environ Sci Technol 28(1):89–117

US EPA (2003). Toxicological review of hydrogen sulfide. http://www.epa.gov/iris/toxreviews/006 1tr.pdf

Chapter 5
Environmentally Benign Pulping Processes

Abstract The pulp and paper industry has been severely impacted by economic conditions and environmental pressures, and it is under intense pressure to improve performance in relation to the release of pollutants. Pressure to adopt new, environmentally friendly technologies is growing due to growing environmental concerns. The general steps of pulping, bleaching, and paper production are all included in an integrated pulp and paper process. Over the past two decades, new environmental laws and consumer activism have drastically changed how chemical pulps are produced. The pulping processes that can address the environmental challenges of the pulp and paper industry are presented in this chapter. These include Steam explosion pulping; High yield pulping; Ozone for high yield pulping; Oxygen delignification; Ozone delignification; Organosolv Pulping; Ionic liquids; Deep eutectic solvents and Biopulping.

Keywords Pulp and paper industry · Steam explosion pulping · High yield pulping · Ozone for high yield pulping · Oxygen delignification · Ozone delignification · Organosolv pulping · Ionic liquids · Deep eutectic solvents · Biopulping · Environmentally friendly technologies

One of the main industries that present a threat to the environment is the pulp and paper sector. The pulp and paper industry has been severely impacted by economic conditions and environmental pressures, and it is under intense pressure to improve performance in relation to the release of pollutants. Pressure to adopt new, environmentally friendly technologies is growing due to growing environmental concerns. The general steps of pulping, bleaching, and paper production are all included in an integrated pulp and paper process. Over the past two decades, new environmental laws and consumer activism have drastically changed how chemical pulps are produced. The pulping processes that can address the environmental challenges of the pulp and paper industry are presented in this chapter.

5.1   Steam Explosion Pulping

High-pressure steam is used to break the bonds between polymeric components and decompression to destroy the structure of lignocellulose. This procedure involves treating the lignocelluloses for a while using high-pressure steam (433–533 K), after which the vessel is quickly depressurized to atmospheric pressure (Agbor et al. 2011). Hemicellulose degrades due to an explosive decompression and a high temperature, and is then extracted as a water-soluble fraction. At mild reaction conditions, cellulose remains mostly intact with little depolymerization. By cleaving its β-O-4 linkages, lignin depolymerizes, and the fragments that remain condense to create a more stable polymer. The lignocellulose structure is broken down by the adiabatic expansion of steam within the cell wall, which causes component redistribution.

This process causes the hemicellulose to break down and the high tempera-ture converts the lignin, increasing the potential for cellulose hydrolysis increasing the potential for cellulose hydrolysis. Enzymatic hydrolysis was 90% in 24 h for poplar chips pretreated by steam explosion, in comparison to only 15% hydrol-ysis for untreated wood chips. Several factors have been discovered that influence the pretreatment of steam explosions. These are residence time, moisture content, temperature and chip size. Optimal solubilization and hydrolysis of hemicelluloses can be achieved using higher temperature and shorter residence time (270 °C, 1 min) or lower temperature and longer residence time (190 °C, 10 min). High-pressure steam is used in steam explosion to break the bonds between polymer components and depressurize to destroy the structure of lignocelluloses.

Lower temperatures and longer residence times are more promising, according to studies, as they prevent the formation of sugar breakdown products that prevent further fermentation. The majority of the carbohydrates during pretreatment get solubilized in the liquid phase by hemicellulose, while lignin undergoes transfor-mation at high temperatures. The solid fraction's cellulose becomes easier to access, improving the lignocellulosic feedstock's digestibility. The process of hemicellulose being broken down into its constituent parts, glucose and xylose, by the action of acetic acid is referred to as "autohydrolysis." It is thought that the acetyl groups connected to hemicellulose and other acids released during pretreatment are what produce acetic acid.

When used as a catalyst, carbon dioxide, sulphur dioxide, or sulfuric acid can significantly improve steam explosion. The recovery of hemicellulosic sugars is increased, the generation of inhibitory compounds is reduced, and the enzymatic hydrolysis of the solid residue is improved when an acid catalyst is used, according to research. Steam explosion has proven to be very effective, for the pretreatment of hardwoods and agricultural residues. However, it is not so effective for softwoods. The use of an acid catalyst in this situation is crucial.

The NREL pilot plant at SEKAB in Sweden, the Italian steam explosion program at Trisaia in southern Italy, and the demonstration-scale cellulosic ethanol plant at Iogen in Canada are all being investigated as this pilot-scale process approaches commercialization. The Mesonite process for manufacturing fiberboard and other

Table 5.1 Benefits of steam explosion

Uses few chemicals
Does not cause an excessive dilution of the produced sugars
Uses little energy, and there is no recycling or cost to the environment

Table 5.2 Drawbacks of steam explosion

The risk of condensation and precipitation of soluble lignin components due to incomplete destruction of the lignin-carbohydrate matrix makes the biomass less digestible
At higher temperatures, there is destruction of a portion of the xylan in hemicellulose and the potential production of fermentation inhibitors
Pretreated biomass must be water washed to remove the inhibitory materials as well as water-soluble hemicellulose due to the formation of degradation products that are inhibitory to microbial growth, enzymatic hydrolysis, and fermentation. This lowers overall saccharification yields because soluble sugars, like those produced by hemicellulose hydrolysis, are removed

products has commercialized the use of non-catalytic steam pretreatment processes (Agbor et al. 2011; Mosier et al. 2005a, b; Sun and Cheng 2002; Weil et al. 1997; Wright 1998).

Tables 5.1 and 5.2 show the benefits and drawbacks of steam explosion.

5.2 High Yield Pulping

High yield pulp (HYP) is made using mechanical, chemical, or combined mechanical and mechanical unit processes from various pulpwoods or annual plants, with a yield of over 85%. HYP includes alkaline peroxide mechanical pulp (APMP), bleached chemi-thermomechanical pulp (BCTMP), preconditioning refiner chemical alkaline peroxide mechanical pulp (P-RC APMP) (Bajpai 2015). Depending upon the type of wood used, the brightness levels, physical strength characteristics and the properties of HYP produced by diverse pulping methods are generally comparable. HYP is a more affordable alternative to hardwood bleached kraft pulp when making writing and printing paper (Ford and Sharman 1996; Reis 2001; Xu 2001; Zhou 2004; Zhou et al. 2005).

Fine papers can benefit from hardwood HYP's increased bulk, opacity, stiffness, and printability (Johnson and Bird 1991). Papermakers can use less pulp for producing paper with the same calliper and stiffness by using high-bulk HYP. Currently, 10% of HYP pulp is used by paper manufacturers to make wood-free paper grades. Most writing and printing paper, including coated and uncoated papers, could use up to 50% HYP pulp. HYP differs significantly from bleached kraft pulp in terms of its qualities. It is inexpensive and produces higher pulp yield. The majority of the lignin from the wood is still present in the fibres of HYP pulp, making the fibres relatively stiff and resistant to lumen collapse. However, kraft pulp, which is essentially lignin-free, has fibres that have collapsed and are very flexible.

In comparison to bleached kraft pulp fibres, HYP shows higher bulk and usually poorer bonding properties due to their rigid tube-like structure. The refining process of HYP pulping results in a wide fiber size distribution because of peeling and delamination along the fiber length and breakage of some fibers (Bajpai 2015). HYP pulp has a significantly higher specific area than bleached kraft pulp because of the presence of fibre fibrils and fragments, and fibrillated fibres. This may have an impact on some papermaking processes, such as retention of fillers, internal sizing, and polymer adsorption. Additionally, bleaching with alkaline hydrogen peroxide used in the production of HYP significantly alters the chemical and physical characteristics of fibres (reduced bulk, opacity, improved bonding).

HYP has more anionic groups, like carboxyl and sulfonic groups, than kraft, resulting in a high fiber charge density than bleached kraft. Additionally, HYP has very high anionic trash content in comparison to bleached kraft pulp. When HYP is used in place of bleached kraft pulp in the papermaking process, these particular chemical and physical characteristics of HYP may have an impact on the wet-end chemistry. Kraft pulping is the predominant pulping method used today all over the world. However, the production of HYP has increased quickly in China. Hardwoods are the main raw materials used to make HYP in China (Bräuer et al. 2012; Jerschefske 2012; Yanhong et al. 2015). The largest producer of HYP in world is China. Both the BCTMP process and PRC-APMP process are used. The majority of the recently installed production lines are using the PRC-APMP process. This type of pulp has several well-known benefits, including the lower production costs, higher opacity, and effective formation of paper. In terms of cutting-edge technologies, China's HYP, dominated by the PRC-APMP process, has a significantly high energy input than the kraft process but significantly less wood consumption. Emissions of carbon dioxide from HYP production are likely to be significantly lower than those of kraft process when the saved wood in forests and plantations is taken into account as an increase in carbon storage.

Based upon the idea that wood saved from the production of HYP (versus kraft pulping) is regarded as carbon storage facility, HYP offers significant environmental advantages; has lower carbon footprint in comparison to kraft pulp. A new HYP with this purpose was developed with the help of researchers from FP Innovations and their partners from Canada and China. This team leveraged more than $1 million. Results have shown that high-yield pulp substitution levels on the market today can be increased in paper furnish by up to 40% without having a materially negative effect on the physical strength properties of paper and printability. Since three years ago, the use of HYP has been constrained due to worries about the final product's quality and paper machine operations.

HYP has increased interest in producing high quality fine paper grades offering reduced raw material costs, improved printability, and increased opacity and paper bulk (Holmbom et al. 1991; Levlin 1990; Pan 2001; Gullichsen et al. 2000). HYP is also environmentally friendly as it has a higher yield during the manufacturing process and a lower carbon footprint in comparison to kraft pulp. This results in fewer trees and chemical inputs per tonne of paper, improving cost-competitiveness and lowering environmental impact overall. Researchers are currently looking into ways to enhance some of HYP's characteristics, create better coating and calendaring techniques, and lessen or completely reduce HYP's detrimental impact on wet-end chemistry.

5.3 Ozone for High Yield Pulping

Despite the ongoing rise in manufacturing costs, including the price of energy and fibrous materials, production of superior quality pulp and paper through an energy-efficient pulping process has always been the main focus for maintaining competitiveness. The presence of lignin in mechanical pulps is one of the major causes of reduced interfibre bonding, even though the hydrophobic lignin must be preserved for imparting mechanical pulp its higher yield and bulk properties (Li et al. 2010). The material richer in lignin on the surface of the fibre can be chemically altered to enhance mechanical pulp fibre bonding while retaining higher yield. Ozone is a potent oxidizer and has already been applied in the bleaching of industrial pulp.

It has a favourable impact on pulp qualities as well as energy savings. Several researchers have studied the use of ozone for high yield pulping. Ozone was first studied to reduce the energy consumption and enhance the strength properties of thermomechanical pulp (TMP) gathered from the main line or reject line, (Allison 1979, 1980). The first time ozone was utilised in a mechanical pulping procedure in 1964 (Ruffini 1966). After secondary refining, it was primarily used on SGW, RMP, or TMP pulps to improve the strength properties. Efficacy can be categorized as TMP > RMP > SGW. It has also been examined during the defibration and beating of the TMP or CTMP process for reducing energy requirement and improving the pulp quality (Allison 1979; Vasudevan et al. 1987).

Ozone serves as a refining agent during pulp refining, supplementing the mechanical energy typically provided by refiners. Ozone's softening effect makes fibre separation easier. For achieving commercial implementation, Hostachy (2010a, b, c, d) made an effort to explain the remaining restrictions. The two key components were the development of the full-scale mixing apparatus in terms of practical considerations and identifying the best application point for application for maximizing the reduction in energy. Use of ozone at a dose of 20 kg/ton on either Spruce or Pine TMP pulps in a thermo-mechanical process can save about 0.65 MWh per tonne of pulp while maintaining the same final freeness. When energy for the production of oxygen and ozone is taken into account, the net savings per tonne of pulp are approximately 0.4 MWh. (Lindholm 1977a, b, c).

For softwoods, brightness was found to be reduced, while for hardwoods, the opposite was true. This demonstrated that one crucial factor in the ozone reaction was the lignin structure.

Intermediate application of 2.5% ozone to primary refined birch CTMP in an alkaline medium reduced energy requirement by 45% and increased tear and tensile properties by approximately 90%. But, pulp yield and brightness decreased by 3 points. Oxidation of lignin with ozone changes the fiber surface.

A thorough investigation into the impact of interstage ozone treatment was made. With a 3% ozone charge, up to 40% energy savings were possible, and the pulp quality improved with both hardwoods as well as softwoods without degrading the brightness or pulp bleachability (Kojima and Yoon 1991; Petit-Conil et al. 1998; Soteland 1982).

At the fibre surface, components of the fibre wall reacted with ozone. By opening the aromatic ring, oxidising the lateral chains linked to lignin's depolymerization, and forming water-soluble organic acids, ozone reacted with lignin. The composite lamella underwent this delignification, which changed the flexibility of the fibres. It was discovered that polysaccharides can generate some aldonic acids by oxidation of terminal hemiacetal groups. By cleaving the pyranoside ring, oxidation of primary and secondary alcoholic groups generated several carboxyl and carbonyl groups. Hemicellulose in mechanical pulp was affected, and cellulose was found to be normally shielded by lignin (Magara et al. 1998). These processes changed the hydrophilic character of the fiber surface. The potential for interfibre bonding increased as a result. Ozone may also be used to refine chemicals. It promoted microfibril separation and increased fibre wall hydration. After ozonation, this led to a decrease in pulp freeness. Lignin in ozonated softwood TMP was analyzed by 13C-NMR (Robert et al. 1999). Lignin extracted from the chemical pulps showed significantly more structural modifications than the carbon skeleton. No correlation was found between increased lignin carbonyl groups and increased strength. The polymeric structure of the lignin in mechanical pulp was not degraded by ozone. Increased fiber flexibility has been used to explain improved pulp strength properties.

By ozonating mechanical pulp, multilayer board's bulk and internal cohesion can be improved. Long fibres' Scott bonds were increased by 250% at low consistency with a charge of ozone that was less than 3% without losing bulk. In CTMP effluents, ozone selectively removed fatty acids and resins. Small ozone charges are sufficient to break down these dangerous elements in effluents and lessen toxicity. The cost of energy has significantly increased in recent years, which has an impact on mechanical pulp producers. Ozone production requires electricity, but using it in the process could reduce specific energy requirements and also enhances pulping, bleachability, pulp quality and paper machine runnability (Roy-Arcand and Archibald 1996).

To lower energy consumption in TMP pulping, Lecourt et al. (2007) studied ozone in the main line and in the reject line. Using ozone in the primary or secondary refiners altered the mechanisms for separating fibres; the requirement of electrical energy reduced by 10–20%. Ozone primarily oxidised wood extractives, which changed the quality of final pulp and chemistry of the fibre surface. Treating TMP rejects with ozone appears to be the extremely promising technology. A 1–2% ozone charge reduced energy use by 10–20%. Additionally, it led to lignin surface modification and extractive degradation of wood. The final pulp's strength characteristics and peroxide bleachability were enhanced by reintroducing these ozonated refined reject fibres.

These benefits could more than offset the expense of producing and using ozone.

According to Sun et al. (2013), the mechanical pulp treatment with ozone depends on the presence of sodium hydroxide. Ozone performs well when the pH is in the range of 5.4–6.15 and the reaction is conducted with whole primary pulp, The surfaces of short fibres and fines are much larger as compared to long fibers. Short fibres have more contact with chemical components (Han et al. 2008). Sun et al. (2014a, b) looked at how the primary long-fibre fraction was selectively refined and then subjected to an interstage ozone treatment to determine how this affected energy usage and pulp quality.

To compare with the conventional TMP process, the secondary pulp was again combined with the primary shorter fiber fraction to form the starting pulp. Regarding the refining of energy, selective refining shows important advantages. When compared to the control TMP trial, a pulp with 100 mL of freeness could save about 15% of the overall refining energy. Another 13.8% in energy can be saved by using 1.5% ozone. The handsheet's physical characteristics are not significantly altered by selective refining, but there is a noticeable rise in the light scattering coefficient. In contrast to selective refining alone, treating the primary long fibres with ozone may change the properties of secondary pulp. However, if primary short fibres are again combined with secondary pulp, this improvement might be lost. This is explained by the low binding capacity of the primary fines.

The same phenomenon does not exist in the optical properties. The recombination of long and fine fibers increases the opacity, brightness, and light scattering coefficient of the handsheet.

A promising new application area has opened for using ozone in high resolution processes. A sustainable supply of raw materials and better management of energy demand are the main concerns of the pulp and paper industry.

5.4 Oxygen Delignification

Oxygen delignification (ODL) of pulp is a tried-and-true technology that is being used for more than 40 years. Due to the industry's trend toward ECF bleaching and lower emissions of organochlorine compounds, oxygen delignification has become a crucial process (Bajpai 2018a). Oxygen can successfully delignify Kraft pulps made from hardwoods and softwoods, sulfite pulps made from hardwoods and softwoods, nonwood pulps, and other pulps. These pulps are used to make a variety of products, including tissue paper, copy paper, grease-proof paper, writing and printing papers, paper board and newsprint.

The "oxygen stage" was established when significant concerns about effluent discharges from bleach plants, energy use, pollution control began to emerge in the late 1960s. These concerns were initially focused on decreases in BOD.

The most growth in ODL occurred in Sweden in the 1970s and Japan in the 1980s in an effort to lower the cost of bleaching agents. Due to the problem with chlorinated organics, the late 1980s witnessed a significant growth. Prior to the bleach plant, a significant reduction in lignin is required for TCF production. The decision to use ODL is based on the mill's specific economic, technical, and environmental requirements. The following are the main benefits of ODL with modified bleaching over traditional bleaching: Retention and reuse of the chemicals and organics that were extracted for use in the oxygen stage; combustion of the recycled organics to produce energy; partially replacing chemicals based on chlorine, mainly chlorine gas, for the production of ECF pulp; elimination of all chemicals based on chlorine in TCF sequences; energy efficiency.

Only 12.5% of the energy of chlorine dioxide, or equivalent chlorine is required to produce oxygen (Tench and Harper 1987; Pikka et al. 2000; Gullichsen 2000). Despite the growing interest in reducing environmental pollutants, the majority of mills use ODL to help with the economics of modernizing or expanding bleach plants (Enz and Emmerling 1987; Tatsuishi et al. 1987). The primary objective of ODL in some nations is to reduce the generation of organochlorine compounds, particularly chlorinated phenols, in order to reduce the biological effects on the environment caused by the bleaching effluent. ODL can possibly shorten the bleaching process and greatly improve the efficiency of bleaching process, provided that an effective washing step is carried out following the oxygen stage.

As the amount of lignin entering the bleach plant is less, there is a substantial reduction in bleach chemicals requirement and a reduction in cost due to oxygen's lower cost than hydrogen peroxide and chlorine dioxide. After kraft pulping, ODL is used as the first bleaching stage for reducing the kappa number by 45–65%. The ODL stage effluent can be recovered and sent to the recovery boiler by incorporating it into the brown stock washing system. ODL method delignifies the pulp more selectively. However, the ODL stage cannot be extended to very low kappa numbers because ODL also reduces pulp strength and viscosity.

The integration of the ODL stage into the brown stock washing system is shown in Fig. 5.1. After cooking, the pulp is washed for reducing the amount of dissolved cooking material that enters the ODL stage. Alkali and oxygen are combined with the pulp. The pulp is discharged to a blow tank following delignification and bleaching in the pressurized reactor. Steam is added to raise the temperature to 90–110 °C after separation of gases there. After ODL and separation of gas, the pulp is subjected to washing for recovering the dissolved organic matter and used chemicals. As a result, between 2 and 4% of the pulp yield is lost. Low selectivity, low efficiency, and low reactivity are some of oxygen's oxidizing properties.

It has a low environmental impact and a moderate capacity to bleach shives and other particulate matter. For ODL reactions to take place, high temperatures (85–115 °C) and pressures (4–8 bars) are required due to the low reactivity. The moderate selectivity of oxygen can occasionally result in a significant loss of pulp viscosity, necessitating the addition of a viscosity protector like magnesium sulfate to the pulp in the case of softwoods. The oxygen delignification system must be supplemented

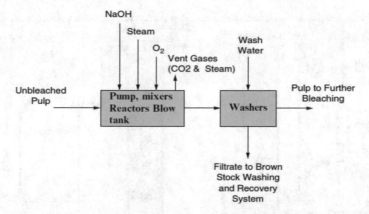

Fig. 5.1 Incorporation of the oxygen delignification stage in brownstock washing and cooking liquor recovery cycle (Based on Gullichsen 2000; McDonough 1996)

with significant quantities of the reagent due to its low efficiency (15–25 kg oxygen per tonne of pulp).

ODL is performed commercially using two kinds of systems. These are typically categorized as systems with medium or high consistency. Since the beginning of the 1990s, mill installations have been dominated by the medium consistency (MC) system because of its improved selectivity and lower investment costs (European Commission 2001). However, installations with a high consistency (25–30%) are also utilized. The OXYTRAC system is one modern two-stage ODL process system (Bokström and Nordén 1998).

Figure 5.2 shows flowsheet of typical medium-consistency oxygen delignification, Fig. 5.3 shows equipment of medium consistency oxygen delignification and Fig. 5.4 shows typical Oxytrac system.

To get the most pulp out of the pulp, the standard is to aim for a higher kappa number in the digester and a higher reduction of kappa number in the oxygen stage. A two-stage process with or without intermediate washing for medium consistency is preferred by the majority of new ODL installations over a single stage. The stage's selectivity and effectiveness have been emphasized in ODL advancements (Johnson et al. 2008).

Although ODL requires considerable capital investment to implement, it is more selective than the majority of extended delignification processes (Gullichsen 2000; McDonough 1996).The major advantages of ODL are a reduction in chemical consumption, pollution, and bleaching chemical costs. The kraft chemical recovery system is compatible with both the materials extracted from the pulp and the chemicals that are applied to it. This makes it possible to recycle effluent from the ODL stage through the brown stock washers to the recovery system, lowering the potential environmental effects of the bleach plant. The reduction is generally proportional to the delignification level of the ODL stage.

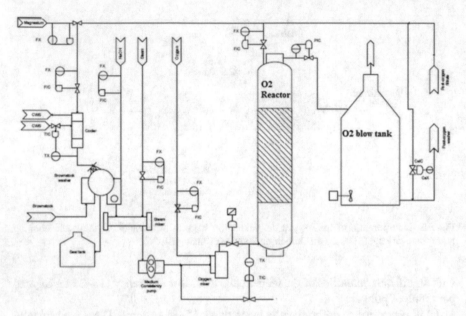

Fig. 5.2 Flowsheet of typical medium-consistency oxygen delignification

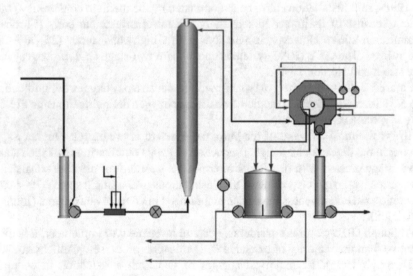

Fig. 5.3 Equipment of medium consistency oxygen delignification

This is true not only for organochlorine compounds, but also for COD, BOD and color that are linked to effluents from bleach plants. However, due to the removal of lignin during the oxygen stage, the color decrease is greater than anticipated. Prior to chlorination, oxygen delignification lowers the kappa number, which reduces the effluent load from bleach plants.

Fig. 5.4 Typical Oxytrac
system (Based on Alejandro
and Saldivia 2003; Bokström
and Nordén 1998)

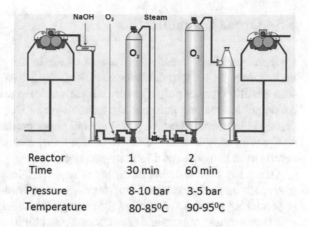

Reactor	1	2
Time	30 min	60 min
Pressure	8-10 bar	3-5 bar
Temperature	80-85⁰C	90-95⁰C

In recent years, the use of oxygen bleaching in industry has grown extremely quickly. There are now more process options available when using medium-consistency equipment which is why mills in North America are more interested in the technology today. This is primarily due to increased environmental concerns. Despite its advantages over oxygen, ozone will likely only be utilized in conjunction with oxygen.

Most kraft plants utilizing ODL utilize oxidized white liquor as the alkali source for maintaining the sodium/sulphur balance in the chemical cycle (Colodette et al. 1990a, b). White liquor oxidation typically utilizes air systems due to their lower operating costs. The strength properties of oxygen bleached pulp are comparable to pulps bleached by traditional method, despite the lower average viscosity of oxygen bleached pulp.

Modified cooking and oxygen delignification are always planned into modern mills. Both approaches must be taken into account in conjunction in order to account for the effects on the environment (COD and AOX discharges). At a given breaking length, the burst and tear factor do not differ significantly. In addition to lignin, oxygen reacts with other pulp materials resulting in numerous byproducts and trace elements that could have an impact on the process, making this process difficult. The two-reactor method, which makes use of separate reactors with different pressures and temperatures, is used when higher delignification rates are needed. The consistency of the pulp, the presence or absence of catalysts or additives, the initial kappa number of the pulp, pH, temperature, presence of transitional metals and the amount of alkali, are just a few of the many distinct factors that influence the degree of delignification.

The greatest benefit of ODL is the reduction of pollutant emissions (Bajpai 2010, 2012, 2015). New developments aimed at lowering wastewater emissions from bleach plants are compatible with the technology. Subsequent washing with oxygen should provide the bleaching plant with well-washed pulp to avoid carry-over of dissolved organic and inorganic substances.

5.5 Ozone Delignification

Over the past three decades, the use of ozone in the pulp and paper industry has undergone a tremendous development (Hostachy 2010a, b, c). Ozone bleaches more than 8 million tons of pulp annually with advances in reliable and affordable on-site ozone production and pulp mixing technology. In fact, ozone holds a very special place in the overall chemistry of pulp and paper production. Ozone is a gas made solely from oxygen and is a "superoxidant" that should be used immediately as it reverts back to oxygen as a final by-product.

Ozone has been tested on a variety of materials, including virgin and secondary fibers, sludge, wastewaters from processes, and wastewaters from the pulp and papermaking process. Its reaction times range from seconds to minutes.

Manufacturing of pulp and paper is renowned for its complexity when it comes to the materials, processes, chemicals, and contaminants used throughout the production chain. In the pulp and paper industries, ozone has been used in wastewater treatment methods and paper pulp delignification technologies (Mamleeva et al. 2022). Ozone is a potent oxidant that doesn't produce any AOX, making it a potential substitute for oxygen delignification. Ozone replacement for chlorine dioxide has shown to be very effective in lowering chlorine dioxide load.

According to Simões and Castro (1999), ozone selectivity during delignification of kraft pulp is only moderately affected by temperature and ozone concentration. This suggests that delignification and cellulose degradation share similar activation energies.

Chirat et al. (2005) studied the use of ozone in place of or in addition to oxygen on high kappa pulps to increase overall pulp yield. Softwood kraft pulp of kappa numbers ranging from 27 to 60 underwent ozonation before being completely bleached with DEDED. When both pulps were bleached using DEDED sequence, the pulp of 60 kappa number showed a yield that was about 2% higher than the 27 kappa number pulp. It also showed 1% better yield than the 60 kappa number pulp treated by ODEDED.

The greater concentration of lignin, which stops carbohydrates from degrading, would explain the greater retention of carbohydrates. The same pattern was seen in case of the hardwood kraft pulp. The measurement of pulp strength properties revealed that the pulps of higher kappa number treated with ozone were much easier to refine in a PFI mill and displayed identical or better strength properties in comparison to reference pulps. In another set of experiment, ozone and oxygen were used to first treat high kappa softwood kraft pulp, after which O(ZD)EpDED and OODEpDED were compared. Even though less ozone was introduced, a benefit in yield could still be measured.

It is possible to prebleach without the use of elemental chlorine by adding a small amount of chlorine dioxide prior to ozonation (Lachenal et al. 1991; Dillner and Tibbling 1991). This improves the selectivity of ozone delignification significantly. A sulfite mill's first stage also employed ozone. Ozone would serve several purposes in this scenario because the pulp is already acidic after cooking (the ozone stage doesn't require the addition of any acids), and as a result, the Z stage effluent could be sent to the recovery boiler. This might address the environmental problems that few mills have (reduction of COD).

Additionally, ozone is highly reactive to extractables. Ozone treatment of sulfite pulps of 30 and 50 kappa number was conducted. The comparison of Z Ep P sequence was made with the control sequence EpP. Application of the EpP sequence to 30 kappa pulp gives a bleaching yield of about 15 at a bleaching yield of 93.3 compared to a Z Ep of 95.1 by the use of 0.5% ozone at the similar kappa number. For high kappa pulp of 50, a yield increase of about 2 points was observed when Z Ep P was used instead of EpP. Ozone has been used by many researchers to maximize lignin removal.

Binder et al. (1980) addressed the delignification of various biomass. They used a diluted wheat straw slurry to handle the reaction with ozone. About 40% of the lignin was removed after treatment for 25 h. These researchers concluded that this process was not a practical pretreatment process.

Miron and Beh-Ghedalia (1982) performed experiments to determine the feasibility of ozonation and other techniques. The straw was treated with sulfur dioxide and 5% sodium hydroxide. The temperature of the reaction was maintained at 70 °C. As a result, delignification by ozone was found to be very low at 22.5%.

Vidal and Molinier (1988) studied ozonation at acidic pH. Acetic acid was used at a dose of 45%. The ozone-oxygen gas stream contained 65 mg/L of ozone. About 66% of the lignin from common sawdust was removed in 360 min, and about 30% of the material was lost during the reaction.

Silverstein et al. (2007) produced ethanol from cotton stalks by spraying ozone gas onto a 10% (w/v) slurry of water and cotton stalks for 30, 60, and 90 min continuously at 40 °C. The ozone reaction did not produce the expected outcomes.

Garcia-Cubero et al. (2009) pretreated wheat straw with ozone at a dose of 2.7% w/w and a flow rate of 60 L/h in a fixed-bed reactor. At 150 min of ozonation, the acid-insoluble lignin was approximately 43%.

Baig et al. (2015) used ozonation technology for delignifying wheat straw for production of biofuel. For ozone pretreatment to be effective, wheat straw needed to be surface treated. A 1% sodium hydroxide soak followed by a neutral pH wash was found to be sufficient to continue reaction with ozone. Around 90% of acid-insoluble lignin could be removed by controlling variables like reaction time, ozone concentration in the ozone-oxygen feed stream, flow rate, water content, and particle size. This procedure is a possible alternative to biomass delignification for the production of biofuels.

5.6 Organosolv Pulping

Organosolv pulping processes use organic solvent/water media to treat lignocellulosic materials with or without a catalyst (Hergert 1998). These processes use less energy and have a lower impact on the environment. The properties of the cellulose, hemicellulose, and lignin fractions that are produced vary based on the specific conditions of the process. A representative organosolv process is schematically shown in Fig. 5.5 (Vaidya et al. 2022).

A wide variety of organic solvents are used in organosolv processes, either as aqueous solutions containing a catalyst (an acid, base, or salt) or in their natural form. Ethanol, methanol, acetone, and ethyl acetate are examples of solvents with low boiling points that can be readily recovered through distillation; in contrast, triethylene glycol and ethylene glycol have high boiling points but permit operation at low pressures. However recovery is difficult.

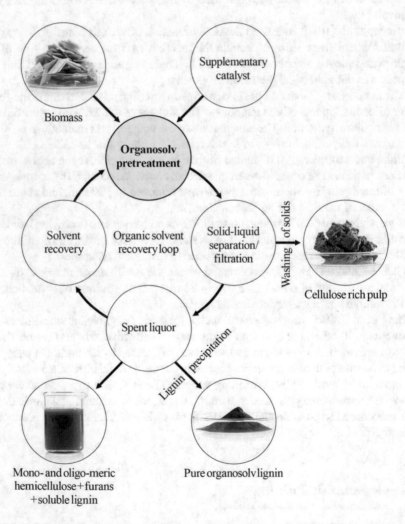

Fig. 5.5 Schematic of a general organosolv process. Vaidya et al. (2022). Distributed under a creative commons attribution 4.0 international license

In any case, all of them allow for efficient delignification of hardwood and soft-wood without jeopardizing the cellulose in the raw material (Rodríguez et al. 2008). In organosolv processes, alcohol is the most commonly used solvent. Ethanol is one of the best options because it is quick to delignify under ideal conditions and easy to recover (Salehi et al. 2015; Oliet 1999). Secondary and tertiary alcohols cannot match the selective delignification offered by primary alcohols. Additionally, hemicellulose losses from methanol are not as noticeable as those from butanol. Sulfur-free tech-nology is likely to reduce emissions of sulfur dioxide and pungent gases for some of these processes. By either directly recovering the solvent after pulping through distillation and combustion of the dissolved wood components or by using it as a chemical feedstock for other products, all new pulping processes aim to create a closed mill. Organosolv pulping's primary objectives are lessening the impact on the environment and increasing the economy of the pulping process.

Table 5.3 shows solvents used in Organosolv pulping and Table 5.4 shows advantages of Organosolv pulping process. Table 5.5 compares Organosolv pulping processes with modified kraft pulping (Young and Akhtar 1998).

Table 5.3 Solvents used in organosolv pulping

Alcohols solvents Methanol, ethanol, n-butanol, amyl alcohol, ethylene glycol, propylene glycol and so on
Organic acids solvent Formic acid, acetic acid, and formic acid + acetic acid, and so on
Ester organic solvent Ethyl acetate
Compound organic solvent Methanol + acetic acid, ethyl acetate + ethanol + acetic acid, and so on
Phenol organic solvents Phenol, cresol and mixed cresol
Active organic solvents Dimethyl sulfoxide, dioxane, diethanol amine, and so on

Bajpai (2021). Reproduced with permission

Table 5.4 Advantages of organosolv process

Use either low boiling solvents (for example, methanol, ethanol, acetone), which can be easily recovered by distillation, or high boiling solvents (for example, ethylene glycol, ethanolamine), which can be used at a low pressure. Thus, it is possible to use the equipment used in the classic processes, for example, the soda and kraft processes, hence saving capital costs
Pulps with properties such as high yield low residual lignin content high brightness and strength can be produced
Valuable by products include hemicelluloses and sulfur free lignin fragments. These are useful for the production of lignin based adhesives and other products because of their high purity, low molecular weight, and easily recoverable organic reagents

Bajpai (2021). Reproduced with permission

Table 5.5 Organosolv versus Kraft pulping process

Pulping process	Raw materials	Pulping chemicals	Process parameters	Kappa No before bleaching
ALCELL	Northern Hardwood	Denatured Ethanol/water Mixture	190–200 °C	N/A
ASAM	Softwood Hardwood Annual plants	Sodium sulfite (alkaline) Anthraquinone Methanol	175–185 °C 11–14 bar	13–20
FORMACELL	Softwood Hardwood Annual plants	Acetic acid Formic acid	160–180 °C	2–10
MILOX	Hardwood Annual plants	Formic acid Hydrogen peroxide	60–80/ 90–110 °C	30–35
Modified kraft	Softwood Hardwood Annual plants	Sodium hydroxide Sodium sulfide	155–175 °C 8 bar	10–20

Based on data from Young and Akhtar (1998)

The three Organosolv pulping processes based on organic acid, alcohol/water and mixed—utilize both organic and inorganic pulping chemicals.

5.6.1 Glycols

Ethylene glycol was first applied to spruce (Hergert 1998; Rodríguez et al. 2008). For pulping, higher glycols like propylene, butylene, and others are useful, especially if the raw materials have already been infused with sulfuric acid. Gast and Puls (1985) discovered that aluminum chloride or sulfate could be used as a catalyst to improve the efficiency of ethylene glycol. They also investigated how quickly birch pulp was pulped with glycol. Research has been conducted on the by-products of lignin recovery during pulping with ethylene glycol as well as the two-step poplar fractionation that results in cellulose, hemicellulose, and lignin in one step (Muurinen 2000a, b).

Numerous authors have examined the use of ethylene glycol for pulping pine, birch, aspen, beech, olive tree prunings, vine shoots, bagasse and forest residues; rice straw with diethylene glycol-ethylene glycol, diethylene glycol-ethylene glycol and soda mixtures; and Miscanthus sinensis, Eucalyptus globules, cellulose linters, and wood sawdust with glycols. The glycol delignification method was successfully modified by the use of glycol- acetic acid-water systems with pinewood and aspen which resulted in significant energy savings (Koell and Lenhardt 1987; Demirbas 1998; Zhang et al. 2016a, b; Rutkowski et al. 1993, 1995; Chaudhuri 1996; Jiménez et al. 2004a). The cause of the ethylene glycol losses during pulping was investigated by Surma-Slusarska (1998).

Pulp production has also utilized glycerols (Demirbas 1998; Meighan et al. 2017; Kucuk and Demirbas 1993). Glycerol treatment of Spruce orientalis and Ailanthus altissima wood produced optimal delignification when an alkali is present but it increased cellulose losses. Additionally, mixtures of ethylene glycol and ethanol have been utilized for pulping, as have butanediol and propylene glycol.

5.6.2 Phenols

Phenol is used as a solvent in the Battelle-Genoa process. This is the most well-known method for pulping. It uses hydrochloric acid as a catalyst and a solvent having a higher boiling point. The use of hardwood, spruce, and herbaceous plants makes the procedure particularly effective. However, softwood processing typically takes longer and results in pulp with less desirable properties than Kraft pulp.

The Battelle-Genoa method has the advantages of producing small quantities of black liquor and having a low cost for industrial plants. However, its drawbacks include issues with pulp washing, reagent recovery, and removal of toxic substances from the effluents, which typically contain low concentrations of phenols and cresols (Asiz and Sarkanen 1989).

Wheat straw, wood and Ulex europaeus have all been successfully delignified using the phenol-ethanol method, which is also well-known (Jiménez et al. 1997a, b; Rezayati-Charani et al. 2006; Sano and Shimamoto 1995; Vega et al. 1997; Vega and Bao 1993).

Schweers (1974); Schweers et al. (1972) have also used phenols in the production of pulp.

Cresols have been also utilized as pulping reagents (Sano et al. 1989). The raw material was delignified for two hours at 180 °C with acetic acid at a consistency of 70%.

5.6.3 Esters

Young and Baierl (1985) patented the ester pulping method, in which water, acetic acid (the catalyst) and ethyl acetate (the solvent) are combined for dissolving the hydrolyzed lignin. This method works well with poplar, but not with other hardwood species like red oak and eucalyptus. Both are inappropriate if the goal is to obtain pulp for bleaching later. This method produces poplar pulp with properties that fall somewhere in the middle of those of Kraft pulp and sulfite pulp (Asiz and McDonough 1987). However, according to some researchers, this method effectively delignifies both softwood and hardwood, resulting in pulp with favorable strength properties (Jiménez et al. 1997a, 1997b).

5.6.4 Methanol

There are currently a number of pulping processes that make use of methanol. According to Muurinen (2000a, b), the alkaline sulfite anthraquinone methanol (ASAM) process is regarded as the superior method because it safeguards carbohydrates and encourages the dissolution of lignin.

Kraft, sulfite, and soda pulping all make use of methanol. The ASAM and soda pulping with methanol (Organocell) demonstration plants have been constructed. ASAM pulping is also regarded as a modified sulfite pulping method. Utilizing methanol may be risky due to the chemical's high flammability and toxic nature (Leponiemi 2008).

The ASAM pulping makes use of sodium sulfite, sodium carbonate, and sodium hydroxide as pulping chemicals. The pulping time and temperature are 60 to150 minutes at 175 °C. Beyond 10 bars, the maximum pressure is reached. Controlling the yield of the pulp, hemicellulose content, and optical properties can be accomplished by altering the ratio of sodium bases to sulfite (Patt et al. 1999).

Metso developed a second procedure known as the FreeFiber process (Enqvist et al. 2006).

The process involves impregnating the material with sodium carbonate prior to cooking it in gaseous methanol. During the heat-up stage, the raw material comes into contact with gaseous methanol after being impregnated. The unused liquor is concentrated and removed. The chips are brought up to the reaction temperature as the heated methanol condenses and releases energy.

Methanol is added as needed to maintain the temperature. The pulp is removed from the reaction and cooled. According to Indufor (2007), the pulp's characteristics are said to be appealing despite the fact that the process does not currently provide obvious benefits for the economy.

According to Patt and Kordsachia (1986), methanol and anthraquinone can be beneficial additions to the alkaline sulfite liquor, resulting in improved strength properties, lower lignin content, increased brightness, and increased pulp yield. The produced pulp can be bleached easily. Bleaching results in a yield that is 5% higher than that obtained from kraft pulp.

The pulp yield is 14 to 18% higher than that of kraft pulping, and the inorganic chemical charge is about 70 to 80% sodium sulfite (Patt et al. 1987).

Because there is no condensation of lignin, there is less lignin left over. Bleaching the pulp does not require the use of chlorine. Because of this, strength properties are higher than those of kraft pulps. In addition, the cooking liquor's methanol content can be decreased from 35 to 20% without affecting the pulp's quality. More than 95% of the methanol could be recovered when the digester's pressure was released (herkules.oulu.fi).

Delignification increased to levels below that of kraft or sulfite pulp when ethanol or methanol were added. Paper properties and increased pulp brightness are two benefits of the ASAM process. The ASAM process produces pulp with a lower Kappa number. The yield is higher and the paper strength properties are improved.

Furthermore, according to Patt and Kordsachia (1986), ASAM pulping completely eliminates the unpleasant odor caused by methyl mercaptans, which are produced during kraft pulping.

The Organocell process makes use of methanol, sodium hydroxide, and catalytic amounts of anthraquinone. This process started out as a two-stage process with a slightly acidic first stage and an alkaline second stage. Aqueous methanol was used for the first stage of cooking, which took place at 190 to 195 °C.

The process was tested at the 5 ton/day demonstration plant level in an aqueous methanol-sodium hydroxide solution for the second stage at temperatures up to 180 °C. The blow line kappa number changes a lot when the temperature changes at such high temperatures. However, it has been demonstrated that the process's initial step can be skipped. A one-stage Organocell process with a pulping temperature of 170 °C and a methanol content of 30% by volume in the cooking liquor can be used. Consequently, controlling the one-step procedure is simpler and fibers made without the first stage are stronger than those made with two stages (Leponiemi 2008, 2011; Schroeter and Dahlmann 1991).

5.6.5 Ethanol Pulping

Clean pulping was the primary goal when the ethanol pulping process was initially developed. The ALCELL pulp production process was developed from this process. This process uses ethanol and water as the cooking medium. There are three distinct steps in the process: lignin extraction for pulp production; recovery of liquor and lignin; and recycling of by-products. The raw materials are cooked for one hour at around 175 to 195 °C in a 50:50 mixture of ethanol and water. Liquor pH is approximately 2 to 3, and the typical ratio of liquid to biomass solids is 4 to 7. At the end of the pulping process, the system uses liquor displacement washing for separating the extracted lignin. This method produces extremely pure, sulfur-free lignin that may have higher value (Bajpai 2018a, b). The only delignifying agent in the ALCELL process is an ethanol solution in water. In a pilot plant that produces 15 tons of unbleached pulp per day, the method has been studied. The pulping medium contained 50% ethanol and water and was heated to 195 °C. Nitrogen was used to keep the pulping vessel slightly overpressured. As a result of the raw materials' deacetylation, the pulping liquor does not require the addition of any acid or alkali (Pye and Lora 1991). Ethanol is recovered by distillation. Hemicelluloses, lignin, furfural and acetic acid, are the primary byproducts of the wheat straw ALCELL process. The pulp still contains the majority of the silica.

According to Winner et al. (1997) approximately 5% of the raw material's silica enters the cooking liquor, the remaining 5% leaves the alcohol recovery system along with the lignin, and the remaining 8% is extracted from the stillage's xylose content.

The ASAE process is an improvement on the ASAM process. Ethanol's higher boiling point makes it possible to pulp at lower pressure. However, in ASAM pulping, there is a greater need for ethanol than there is for methanol. ASAE pulps made from wheat straw have lower levels of lignin, superior strength and good beatability. When compared to kraft pulping, ASAE pulping produces paper with a higher yield and uses less energy (Usta et al. 1999).

Alkaline sulfite AQ-methanol pulping was used to enhance ASAE. Alkaline sulfite AQ-ethanol cooking, on the other hand, requires significantly more ethanol than alkaline sulfite AQ-methanol cooking does. Despite this, ASAE pulp has favorable physical properties—shows good beatability and produces higher yield. When compared to the sulfate method, this method may result in significant energy savings (Azeez 2018).

An alkaline solvent pulping procedure known as the IDE-process consists of three steps that follow one another. They are: Impregnation (I); Depolymerization (D); Extraction (E).The recovery is easier than with traditional alkaline pulping processes because there is no use of sulfur or sodium hydroxide. The IDE process can deal with a variety of raw materials. There would be virtually no effluent from the IDE-pulp mill if the final pulp were bleached in a nearby kraft pulp mill (Hultholm et al. 1995).

In the Punec pulping process, ethanol, caustic soda and anthraquinone, are used. The raw material is delignified in a high pressure digester after first being treated with aqueous alcohol. Acidification is used to separate the lignin from the flash liquor after the lignin and hemicellulose get dissolved in the pulping liquor and flashed into the tank. The liquid rich in hemicellulose is distilled to extract any alcohol that is still present. In the end, the hemicelluloses' aqueous fraction is processed anaerobically to make biogas, fertilizer, or livestock feed.

The process has been tried out in a 4 tons/day demonstration plant. However, neither the specific conditions of the process nor the quality of the pulp are known. Khanolkar (1998) claims that the procedure is pollutant-free, making it worthy of further investigation.

Table 5.6 shows the properties of pulp produced with different alcohol pulping methods.

5.6.6 Organic Acid Pulping

In most acidic pulping techniques, formic acid and acetic acid are used. The temperature and pressure of formic acid pulping are lower than those of acetic acid or alcohol pulping. During delignification, formic acid and acetic acid form the corresponding esters when they combine with lignocellulosic raw materials. Acetic acid or formic acid is also produced when lignocellulosic raw materials are processed in an acidic way. The use of formic and acetic acids has this advantage. However, process equipment suffers from corrosion caused by organic acids, particularly formic acid (Rousu et al. 2002; Saake et al. 1995).

Milox process makes use of peroxyformic acid or peroxyacetic acid. Utilizing hydrogen peroxide, formic acid, or acetic acid, these chemicals are produced on the spot. Performing the process in multiple stages can reduce the amount of hydrogen

Table 5.6 Properties of pulp produced with different alcohol pulping methods

Process	ALCELL	ALCELL	ALCELL	IDE	ASAM	ASAE	ASAE	ASAE
Raw material	Reed	Straw	Kenaf	Reed	Bagasse	Wheat straw	Wheat straw	Wheat straw
Na$_2$SO$_3$ charge %					16–18	12.0	14	16
Alkali ratio					0.7			
NaOH charge %				Na$_2$CO$_3$		3	3.5	4
AQ %				0.022		0.1	0.1	0.1
Methanol / Ethanol %	~ 50	~ 51	60	50%	15–20 *	50	50	50
Cooking temperature °C	195		200			170	170	170
Kappa number	22–24	27–32	~ 30	~ 21	3–6	17.4	0.16.4	16.4
Screened yield %	48–53	51–53	~ 60	38–46	61–63	55.5	56.1	54.4
SR number						25	25	25
Tensile index N.m/g				50				
Burst index kPa.m^2/g			4.4			2.43	3.24	2.99
Tear index mN.m^2/g			9.9	7.8		5.9	7.3	6.1
Brightness % ISO					49–62			

Based on Indufor (2007); Leponiemi (2008); Usta et al. (1999); Winner et al. (1997)

peroxide used. In the middle stage of the three-stage Milox process, the pulp is treated with acetic or formic acid without peroxide. Delignification is higher in the two-stage peroxyacetic acid process than it is in the three-stage peroxyformic process. Sulfur is not used in the Milox process. It is possible to bleach without the use of chlorine chemicals. Numerous issues arise from the Milox process's recovery of cooking chemicals. There is some production of formic acid and acetic acid. Extractive distillation is used to separate the mixture of acetic acid, formic acid, and water. It has been suggested that for the distillation butyric acid be used as a solvent. The Milox process has the potential to become more cost-effective if a mixture of formic acid and acetic acid can be utilized as a solvent (Bajpai 2012; Muurinen et al. 1993; Poppius-Levlin 1991).

Chempolis process is based on formic acid pulping. This process involves only one step, and the cooking time is 20–40 min at 110–125 °C. Acetic acid, is produced in smaller amounts during the process, can also be used in the process alongside formic acid. The pulp is cleaned and pressed with formic acid in two to six stages following pulping. The final washing step uses performic acid, which produces a pulp-like product. The pulp is washed with water for removing the acid. During the acid washing stage, hemicelluloses, lignin and fatty acids are removed from the pulp. Alkaline peroxide (charge 36.5% hydrogen peroxide) is used to bleach the unbleached pulp at the end. The remainder has been made up of water, and the proportions of formic acid to acetic acid have varied from 80/15 to 40/40. Bleached pulp yield is 39%. It is possible to evaporate spent cooking liquor down to 90% dry solids without experiencing viscosity issues as silica does not get dissolved in pulping liquor. After it has evaporated, the wasted liquor can be burned. Furfural, acetic acid and formic acid are all produced during evaporation. Because formic acid is made, it is said that less makeup formic acid is needed. Evaporation condensates can be separated from volatile compounds like furfural and acetic acid through distillation (Anttila et al. 2006; Indufor 2007; Rousu and Rousu 2000; Rousu et al. 2003).

One of the acetic acid pulping methods is the acetosolv method. As a catalyst, hydrochloric acid is utilized. There is no pressure in this process. The Acetocell process, which was developed using the Acetosolv concept, is used to obtain dissolved lignin and furfural from waste liquors. Cooking takes place at a temperature of 110 °C. Acetic acid is used to pulp the material at a high temperature without the use of a catalyst. At high pulp yield, the delignification reduces kappa numbers.

From the Acetocell process, the Formacell process was developed. It is an organosolv pulping method in which a mixture of 10% formic acid and 75% acetic acid is used as the pulping ingredient. The remainder is water. After being cooked at temperatures between 160 and 180 °C, the cooked pulp is bleached with ozone and washed with acid. The Formacell process can produce pulps that dissolve and are suitable for use in paper (Leponiemi 2011; Nimz 1989).

The CIMV procedure is a further development of the Formacell strategy. In the Champagne-Ardenne region of France, a small agrofiber mill has been built, which produces 50,000 tons of paper and 50,000 tons of byproducts annually. Using bagasse or straw, the CIMV process produces bleached pulp, xylose, and sulfur-free lignin. Organic acids from waste liquor are recycled by evaporation. In order to precipitate lignins, which are simple to separate, the remaining syrup is treated with water.

This procedure uses water (20%), acetic acid (50%), and formic acid (20%) as pulping chemicals. In a PPP bleaching sequence, bleaching requires 4% peroxide and 12% sodium hydroxide. The process is atmospheric and the pulping temperature is closer to 100 °C. About 32% of bleached pulp is produced (Delmas et al. 2006; Leponiemi 2011).

Table 5.7 shows properties of pulp produced with different organic acids.

5.6.7 Acetone

Acetone has been used alone for delignification or in combinations with ethanol (De Rosa et al. 1997; Ferraz et al. 2000; Jiménez et al. 1998, 2001a, b, 2002b). On previously steamed raw materials, Baeza et al. (1999); Freer et al. (1999) have used

Table 5.7 Properties of pulp produced with different organic acids

Process	CIMV	CIMV	CIMV	Chempolis
Raw material	Bagasse	Rice straw	Rice straw	Wheat straw
Formic acid %	30	30	20	
Acetic acid %	55	50	60	
Water %	15	10	20	
Temperature °C	107	107	107	
Time min	180	120	180	
Kappa number	28.2	34.6	45.8	6.1
Yield %	49.4	47.5	52.9	
SR number	45		45	42
Tensile index N.m/g				53
Burst index kPa.m^2/g	3.21		2.52	2.4
Tear index mN.m^2/g	4.23		4.38	2.35
Brightness % ISO				

Based on Delmas et al. (2006); Indufor (2007); Leponiemi (2008)

acetone alone or in combination with formic acid. Cotton stalks, poplar wood, and eucalyptus were the targets of one method that utilized oxygen in aqueous acetone. The transfer of oxygen was identified as the primary stage of the delignification process, and the effects of temperature, time, and oxygen partial pressure on the rate of delignification were examined (Delpechbarrie and Robert 1993; Neto et al. 1993). Spruce and other kinds of wood were also used in the process (Kunaver et al. 2016).

5.6.8 Ammonia and Amines

Ammonia or amino base processes yield high yields of pulp because they preserve hemicelluloses. As a result, although 1, 6-hexamethylenediamine (HMDA) produces pulp from both hardwood and softwood with a high yield, bleaching the pulp is challenging. Spruce pulp produced using ammonia, acetone, or methylethyl ketone yields more and has better properties than Kraft pulp, with the exception of the tear index. When combined with ethanol, ammonium sulfide and ethanol can be used to produce strong pulp with a high yield and a low lignin content in comparison to softwood, cereal straw, and bagasse pulp with similar properties to that of hardwood. However, the recovery of reagents, the high boiling point of HMDA, condensation of lignin, and the formation of polluting sulfur volatiles pose challenges to processes utilizing ammonia and amino bases (Asiz and Sarkanen 1989).

The extent to which cellulose, hemicellulose, and lignin were removed from poplar chips was determined to be a function of time, pressure, and particularly temperature and solvent concentration when supercritical ammonia-water mixtures were used to process the chips (Bludworth and Knopf 1994).

Wallis (1978) applied ethanolamines to pine and eucalyptus wood in the late 1970s, and monoethanolamine proved to be more effective than diethanolamine and triethanolamine in pulping. Compared to the Kraft process, yields were 11–16% higher, and the resulting pulp had similar properties related to strength.

Wallis (1980) also used ethanolamine to pulp Pinus elliottii, producing pulp that was as strong as Kraft pulp but in higher yields (5–10%).

Spruce and rice straw have also been pulped with ethanolamines (Jiménez et al. 2003; Kubes and Bolker 1978). It was discovered that the rate of lignin degradation increased when ethanolamine was added to the alkaline cooking liquor (Obst and Sanyer 1980).

Lastly, amines have also been used by some researchers to pulp various raw materials like cotton, beech, spruce, olive tree prunings, and jute (Claus et al. 2004; Jahan 2001; Jahan et al. 2001; Jahan and Farouqui 2003, 2000; Jiménez et al. 2002a, b, 2004a; Sarwar et al. 2002).

5.6.9 Other Solvents

Using formamide and dimethylformamide to process bagasse as a raw material, pulps with good physical and mechanical properties were produced. Eucalyptus pulped with dioxane in the presence or absence of hydrochloric acid (catalyst) by some authors and hardwood pulped with the sulfur dioxide ethanol water system by others (Elmasry et al. 1998; Balogh and Curvelo 1998; Machado et al. 1994; Mansour et al. 1996; Mostafa 1994; Rezati-Charani and Rovshandeh 2005; Selim et al. 1996; Westmoreland and Jefcoat 1991).

5.7 Ionic Liquids

Techniques for pretreatment with ionic liquid (IL) are opening up new possibilities for biomass conversion. These are being studied widely as "green solvents" in chemical synthesis, catalysis, and biomass separation with cellulose for a variety of industrial and academic endeavors. They can dissolve a lot of cellulose in a small amount of water. In the modern industrial fiber-making process, for the direct dissolution of cellulose in the commercial Lyocell process, this technology was utilised. Their potential to restore nearly all of the used ILs to their original purity is what makes them so appealing (Fink et al. 2001; Heinze et al. 2005; Pu et al. 2007; Sathitsuksanoh et al. 2012; Socha et al. 2014; Wu et al. 2004, 2011).

There are two main pathways by which ILs convert lignocellulosic carbohydrates into fermentable sugars. Prior to enzymatic hydrolysis, the biomass is treated to improve its efficiency in the first method. The second method involves dissolving the biomass in a solvent to transform the hydrolysis process into a homogeneous one. It is recognized that ILs facilitate more environmentally friendly applications in reactions and separations. Their very low vapor pressure makes it less likely that people will be exposed to them, which is a big advantage over using volatile solvents (Wang et al. 2011).

As shown in Table 5.8, it is becoming more and more common to dissolve a wide range of lignocellulosic biomass with ILs.

The properties of Ionic liquid are shown in Table 5.9. ILs have a number of benefits over conventional volatile organic solvents as cellulose solvents. They are entirely made of ions and have low hydrophobicity, low toxicity, a wide range of combinations of cations and anion, improved electrochemical and thermal stability, higher reaction rates, lower volatility with potential environmental benefits, the property of having low flammability, higher polarities, lower melting points, and negligible vapor pressure (Wasserscheid and Keim 2000; Zavrel et al. 2009).

According to Menon and Rao (2012), there are two main types of ionic liquids: simple salts made up of just cations and anions and those that are in equilibrium. The imidazolium cation can pair with anions like acetate, methanoate, sulfate, bromide, triflate and nitrate in the most common forms. An estimated 109 ILs in the formulation could be created to optimally pretreat specific biomass by combining anions and cations (Wu et al. 2011).

According to Holm and Lassi (2011), the length and pattern of branching of the cation-integrated alkyl groups in ILs can alter their properties. To dissolve bamboo and wood, the ILs 1-ethyl-3-methylimidazolium glycinate (Emim-Gly) and 1-allyl-3-methylimidazolium chloride (Amim-Cl) were created in that order (Muhammad et al. 2011; Zhang et al. 2013a, b). In pretreatment, not all of the properties of ionic liquids make them good solvents. 1-Butyl-3-methylimidazolium chloride (Bmim-Cl) is an example of an IL that is toxic, corrosive, and hygroscopic (Xie et al. 2012). Amim-Cl is one example of a viscous compound with reactive side chains. Additionally, due to their hydrophobic nature, ILs with long akyl chains frequently obstruct enzymes' nonpolar active sites (Ventura et al. 2012).

Others are being investigated as potential solvents due to their favorable properties. Phosphate-based solvents are less toxic, have a lower viscosity, and are more thermally stable than chloride-based solvents (Mora-Pale et al. 2011).

Emim-Ac (1-ethyl-3-methylimidazolium acetate), has been used to achieve positive results. This is advantageous to in situ enzymatic saccharification because of its biocompatibility and enzymatic activity (Li et al. 2011). The potential for ILs to enhance the digestibility of lignocelluloses is being investigated (Zavrel et al. 2009).

ILs based on Imidazolium dissolve biomass effectively. Despite their effectiveness, imidazolium cations have limited industrial applications due to their high price.

The hydrogen and oxygen atoms of cellulose hydroxyl groups contribute to the dissolution of cellulose in ILs by forming electron donor–electron acceptor

Table 5.8 Dissolution of lignocellulosic biomass by ionic liquids

Switchgrass Emim-Ac
Bamboo Emim-Gly
Rice straw Ch-Aa
Kenaf powder Ch-Ac
Cassava pulp Emim-Ac, Dmim-SO$_4$, Emim-DePO$_4$
Poplar Emim-Ac
Pine Amim-Cl
Eucalyptus Emim-Ac
Wheat straw Amim-Cl, Bmim-Ac
Water hyacinth Bmim-Ac
Rice husk Bmim-Cl, Emim-Ac
Spruce Amim-Cl
Bagasse Emim-Ac

Bmim-Ac 1-butyl-3-methylimidazolium acetate
Dmim-SO$_4$ 1,3-dimethylimidazolium methyl sulfate
Emim-DePO$_4$ 1-ethyl-3-methylimidazolium diethyl phosphate
Ch-Aa Cholinium amino acids
Ch-Ac Cholinium acetate
AMIM-Cl: 1-Allyl-3-methylimidazolium chloride
Emim-Ac 1-Ethyl-3-methylimidazolium acetate
Based on Muhammad et al. (2011); Papa et al. (2012); Perez-Pimienta et al. (2013); Weerachanchai et al. (2012)

complexes that interact with the ILs (Feng and Chen 2008). The three-dimensional network of lignocellulosic molecules is disrupted because these ionic liquids compete with them for hydrogen bonding (Moultrop et al. 2005).The hydrogen bonds are broken when ILs and cellulose-OH interact. As a result, the hydrogen bonds that hold the chains of cellulose together break as a result of the interaction in the end.

Through a strong basicity of the hydrogen bond, it has been discovered that ILs with pyridinium or imidazolium cations paired with Cl, CH_3CO_2, HCOO, CF_3SO_3, CF_3CO_2, and R_2PO_4 anions can dissolve cellulose fibers. The primary bonds between cellulose, hemicellulose, and lignin are broken when lignocellulose dissolves in ILs,

Table 5.9 Properties of ionic liquids

Melting point below 100 °C
Nonflammable
Liquid at room temperature
Salts consisting of large cations (mostly organic) and small anions (mostly inorganic), with a low degree of cationic symmetry
Improve antielectrostatic and fire-proof properties of wood
High electrical conductivity, high solvating properties, and wide electrical window
Low volatility and high thermal stability up to temperatures of about 300 °C

Based on Dadi et al. (2007); Mora-Pale et al. (2011); Vancov et al. (2012)

making hydrolytic enzymes more likely to access the substrate. Li et al. (2011) claim that up to 80% of hemicellulose and lignin can be fractionated using the right anti-solvents, such as alcohol, water, and acetone.

Treatment of raw and steam-exploded wheat straw with BMIMCl significantly increased the yield of enzymatic hydrolysis (Liu and Chen 2006). BMIMCl altered the structure of wheat straw by partially solubilizing cellulose and hemicellulose and reducing the crystallinity and degree of polymerization.

The enzymatic hydrolysis yield of bagasse treated with 1, 3-N-methylmorpholine-N-oxide is approximately two times higher in comparison to that of untreated bagasse (Kuo and Lee 2009).

Using antisolvents, soluble cellulose can be quickly precipitated back into form. According to Zhu (2008), recovery cellulose was identical to the original cellulose in terms of polymerization and polydispersity, but it had a noticeably different macro- and microstructure, with less crystallinity in particular.

Ionic liquids are used as selective solvents for lignin and cellulose because they are capable of dissolving a wide range of biomass of varying hardnesses (Cao et al. 2010). Following the biomass's solubilization at 90–130 °C in the solvent and ambient pressure, water is added to precipitate the biomass in this procedure. Washing the precipitate completes the procedure. After being treated with ionic liquids, hemicellulose and lignin retain their original structure. This enables selective extraction of the unchanged lignin as lignin is highly soluble in solvents and cellulose is barely solubilized (Zhang et al. 2012a, b). This can lead to lignin separation without the need for acidic or alkaline reagents or the generation of inhibitors and an increase in the accessibility of cellulose under normal temperature and pressure conditions. These solvents are expensive. However, due to the solvents' low vapor pressure, their recovery is not expensive. In any case, cellulase is irreversibly inactivated in ionic fluid solvents, which makes biomass conversion less effective and raises overall costs (Zhi et al. 2012). This finding demonstrates the necessity of developing solvents in which cellulase and microorganism are active.

In fields like biomass pretreatment and fractionation, ionic liquids have made it possible to use lignocellulosic materials more effectively in new ways. However, putting these potential uses into practice still faces numerous obstacles. The following are examples:

There are several issues which may be addressed. These are–the need for regeneration; a lack of toxicological data and an understanding of fundamental physicochemical properties; the higher cost of ILs; the mode of action on the hemicelluloses and/or lignin content of lignocellulosic biomass; and the production of inhibitors. Additional study and funding are required to address these issues.

5.8 Deep Eutectic Solvents

The use of DESs in the processing of lignocelluloses is growing in importance because they are cleaner solvents than ionic solvents (ILs). Despite the fact that DESs have more advantages than ILs, research on them is still in its infancy, and they are not yet widely utilized in biomass processing. Higher-quality lignin can be extracted from DESs with a purity of over 90% and can be dissolved which constitutes nearly 60% of rice straw's total lignin. It was discovered that DESs have a very low solubility in water (Kumar et al. 2016; Oliveira et al. 2015; Satlewal et al. 2018; Osch et al. 2016; Yoo et al. 2017).

Abbott et al. (2003) were the first to describe the DES concept as a liquid formed when carboxylic acid and a number of quaternary ammonium salts were combined. After that, DESs were made available as inexpensive eutectic mixtures with IL-like properties (Abbott et al. 2004). Eutectic mixtures of hydrogen bonding acceptors and donors are combined to make them (Xu et al. 2017). Due to their simplicity in synthesis, competitive pricing, and general environmental friendliness, DESs are preferred to conventional ILs. The cost of DES components was ten times lower than that of ionic liquid preparation components. The cost of synthesising DES was only 20% more expensive than IL (Gorke et al. 2008; Satlewal et al. 2018; Mbous et al. 2017). It is unclear how the solvent properties of the resulting eutectic mixtures relate to the molecular composition. However, there have been reports of several promising DESs systems.

Table 5.10 shows the properties of some common DES combinations (Loow et al. 2017).

Over the past few years, there have been an increasing number of publications on DESs (Satlewal et al. 2018). Electrochemistry, fermentation and bioindustrial chemistry, the processing of lignocellulosic biomass, fossil fuels, pharmaceuticals, nanomaterials, food and feed industry, separation and metal processing are all potential uses for DESs (Hadj-Kali 2015; Isaifan and Amhamed 2018; Lee 2017; Shishov et al. 2017; Zainal-Abidin et al. 2017).

Since their first description in 2003, DESs have been increasingly used in a variety of fields. They have recently been demonstrated to be utilized as solvents in a variety of reactions, including hydrolysis, transesterification, esterification, and polymerization. Additionally, it has been reported that they have catalytic effects in a variety of reactions (De Santi et al. 2012; Keshavarzipour and Tavakol 2015; Tran et al. 2016;

Table 5.10 Properties of some common DES combinations

HBD	HBA	Molar ratio (HBD:HBA)	Freezing part (°C)	Density (g cm^{-3})	Viscosity (cP)	Surface tension (mN m^{-1})	Conductivity (mS^{-1})	References
Urea	ChCl	2:1	12	1.25	750 (25 °C)	52 (25 °C)	0.75 (25 °C)	Smith et al. (2014) and Zhang et al. (2012a, b)
Ethylene glycol	ChCl	2:1	−12.9	1.12	37 (25 °C)	49 (25 °C)	7.61 (25 °C)	
Glycerol	ChCl	2:1	17.8	1.18	259 (25 °C)	55.8 (25 °C)	1.05 (25 °C)	
CF$_3$CONH$_2$	ChCl	2:1	51	1.342	77 (40 °C)	–	–	
ZnCl$_2$	ChCl	2:1	–	–	85,000 (25 °C)	–	0.06 (42 °C)	
Urea	ZnCl$_2$	3.5:1	9	1.63	11,340 (25 °C)	–	0.18 (42 °C)	
Imidazole	Bu4NBr	7:3	–	–	810 (20 °C)	–	0.24 (20 °C)	
Ethylene glycol	ZnCl$_2$	4:1	–	1.45	–	–	–	Smith et al. (2014)
2,2,2-Trifluoroacetamide	ChCl	1.6:1	Liquid at 25 °C	1.342	77 (40 °C)	35.9 (25 °C)	–	Abo-Hamad et al. (2015)
Acrylic acid	ChCl	1.6:1	Liquid at 25 °C	–	115 (22 °C)	–	–	
Glycerol	Methyltriphenylphosphonium bromide	3:1	−5.55	1.3	–	58.94 (3 °C)	0.062 (25 °C)	
Ethylene glycol	Methyltriphenylphosphonium bromide	4:1	−49.34	1.23	–	51.29 (25 °C)	1.092 (25 °C)	

(continued)

Table 5.10 (continued)

HBD	HBA	Molar ratio (HBD:HBA)	Freezing part (°C)	Density (g cm^{-3})	Viscosity (cP)	Surface tension (mN m^{-1})	Conductivity (mS^{-1})	References
Triethylene glycol	Methyltriphenylphosphonium bromide	5:1	−21	1.19	–	49.58 (25 °C)	–	
Malonic acid	ChCl	1:1	10	–	721 (25 °C)	65.7 (25 °C)	0.55 (25 °C)	Tang and Row (2013) and Zhang et al. (2012a, b)
1.4-Butanediol	ChCl	3:1	−32	1.06	140 (20 °C)	47.17 (25 °C)	1.64 (25 °C)	
Imidazole	ChCl	7:3	56	–	15 (70 °C)	–	12 (60 °C)	
Acetamide	EtNH$_3$Cl	1.5:1	–	1.041	64	–	0.688 (40 °C)	
Urea	EtNH$_3$Cl	1.5:1	–	1.140	128 (40 °C)		0.348 (40 °C)	
Urea	AcChCl	2:1	–	1.206	2214 (40 °C)		0.017 (40 °C)	

Loow et al. (2017). Reproduced with permission

Mondal et al. 2016; Musale and Shukla 2016; Patil et al. 2014; Singh et al. 2011; Wang et al. 2013).

Biocompatibility with biomolecules like proteins, enzymes, nucleic acids, and microorganisms is one of the most important characteristics of DESs whose applications in the biopharma industry for molecular extractions, biotransformation, and bioorganic catalysis have attracted attention in the recent years (Mbous et al. 2017).

The production of biofuels, common chemicals and value-added products, and by dissolving the polysaccharides and lignin in biomass with DESs as opposed to ILs piques worldwide interest (Oliveira et al. 2015).

The use of DESs in the processing of lignocelluloses has been the subject of numerous research papers (Degam 2017; Loow et al. 2017; Lynam et al. 2017; Oliveira et al. 2015; Procentese et al. 2018; Ren et al. 2016; Satlewal et al. 2018; Tang et al. 2017; Ünlü and Takaç 2020; Vigier et al. 2015; Zulkefli et al. 2017).

Within the framework of the idea of valorizing biomass, Ünlü and Takaç (2020) have comprehensively reviewed the application of DESs for treating agro-industrial wastes. Recent advancements and studies utilizing DESs in the processing of biomass as long-term resources has been reported by Elgharbawy et al. (2020). When used to pretreat biomass, DESs reduced cellulose crystallinity and caused lignification.The capacity of DESs to donate and accept electrons and protons was cited as the reason for this (Vigier et al. 2015). By severing the lignocellulosic biomass's inter- and intrahydrogen bonds, this remarkable property of DES made it possible for the biomass to be broken down (Paiva et al. 2014; Xu et al. 2016).

Lactic acid *betaine*-based *NADESs* and lactic acid-ChCl were used to pretreat rice straw (Kumar et al. 2016). Over 90% purity of lignin was obtained. The rice straw was found to have separated about 60% of the lignin. Furthermore, the treatment with lactic acid-ChCl resulted in a higher degree of lignin solubility. The amount of lignin that was extracted increased by approximately 22% when 5% water was added during pretreatment. After the pretreatment, the biomass's crystallinity index decreased, and there were slight structural differences between the crystalline and amorphous cellulosic portions. ChCl-lactic acid was better able to dissolve lignin when the acid content was also raised to a molar ratio of 1:2 (Francisco et al. 2012).

Changing the molar ratio of ChCl-lactic acid from 1:2 to 1:15 enhanced the extractability of corncob lignin from 64.7 to 93.1%. However, not all kinds of carboxylic acids can be used in DES for biomass pretreatment. On the other hand, the interaction between the hydroxyl groups in ChCl–ethylene glycol resulted in the removal of 87.6% of the lignin (Zhang et al. 2016a, b).

At temperatures greater than 90 °C, for instance, the carboxyl groups of malonic and oxalic ChCl–dicarboxylic acid DESs have the potential to activate and release carbon dioxide.

The effectiveness of various DESs in dissolving lignin in biomass is significantly influenced by the DES mixture (Spronsen et al. 2011).

The solvent properties determine the effectiveness of lignin extraction in DESs based on the Kamlet–Taft polarity parameters (basicity or ability to accept hydrogen bonds, acidity or ability to donate hydrogen bonds, polarity and polarizability) (Brandt et al. 2013; Jessop et al. 2012).

However, the delignified cellulosic portion of the biomass may undergo additional valorization processes. In addition, lignin can be utilized in a variety of other processes, including the extraction of chemicals and energy, despite the fact that it cannot be used in the fermentation process (Cherubini 2010).

Lignin, accounts for 15–25% of the dry matter in lignocelluloses and is the largest non-carbohydrate component. Douglas fir softwood and Poplar hardwood yielded lignin with a purity of over 95%. The lignin product preserved the native lignin activity while having a lower molecular weight and better polymer stability. As a result, it proved that it could be used as a useful chemical precursor (Constant et al. 2016; Ragauskas et al. 2014). Compared to ChCl–glycerol, which did not significantly remove lignin, hemicelluloses or cellulose, from the biomass at 145 °C, ChCl–lactic acid removed the most lignin from poplar (58.2%) and Douglas fir (78.5%) (Alvarez-Vasco et al. 2016).

Due to lignin's powerful ability to bind to the cellulose and hemicelluloses in the biomass, DESs were less soluble in lignocelluloses than pure lignin (Zhang et al. 2016a, b). Surface morphology analysis revealed that the pretreatment with DES damaged the supramolecular structure of the biomass, resulting in the development of a porous structure on the surface of biomass, with regard to sugar recovery.

After ChCl–formic acid pretreatment, in untreated corn stover, the initially stiff bundles of fibers became loose and contained a few small particles, which are thought to be lignin aggregates and residual hemicelluloses. The removal of lignin and hemicelluloses made enzymes more likely to find cellulose, which led to a significant recovery of glucose. After enzymatic hydrolysis and pretreatment with ChCl–formic acid, the maximum glucose yield was 99% (Xu et al. 2016).

Certain DESs showed good performance in pretreatments of biomass using ethyl ammonium chloride:ethylene glycol for the pretreatment of oil palm trunk fibers with 74% glucose yield. ChCl:glycerol pretreatment of corncob resulted in a 92% transformation of glucan. A 95% increase in glucan conversion was achieved using ChCl:imidazole. When rice straw was pretreated with choline chloride:urea and choline chloride:oxalic acid, about 90.2% glucose yield was obtained. When corn stover was pretreated with choline chloride: formic acid, 99% yield was obtained (Hou et al. 2017; Procentese et al. 2015; Xu et al. 2016; Zulkefli et al. 2017).

5.9 Biopulping

"Biopulping" refers to the process of treating wood chips with white rot fungi before mechanical or chemical pulping. In "biomechanical" pulping, fungi are used to replace chemicals in the pretreatment of wood for mechanical pulping, cut down on energy use, and improve paper strength. Biopulping makes lesser use of bleaching agents than chemical pulping does either by increasing the cooking capacity, extending the cooking time, or reducing the amount of pulping chemicals used. The effectiveness of delignification boosts pollution reduction and indirectly saves energy for pulping (Kirk 1993; Kirk et al. 1994).

5.9.1 Biomechanical Pulping

5.9.1.1 Microorganisms

In order to enhance the selectivity of lignin degradation and, consequently, the specificity of biopulping, the Swedish Pulp and Paper Research institute put a lot of effort into creating cellulase-deficient mutants of some white rot fungi. In one study conducted with pine and spruce the tensile index increased and energy savings of up to 23% was obtained.

Bagasse has been successful on a large scale but the results with wood chips were not as good. According to Table 5.11, pretreatment of the bagasse with fungi reduced the energy requirement from 4800 to 1700 kWh/t for the production of chemimechanical pulp (CMP) of 70 SR using the Cuba-9 process (treatment with 6% NaOH at 90 °C for 10 to 20 min). Biochemimechanical pulp (BCMP) had superior strength properties to CMP, but the BCMP yield decreased slightly because of the fungal degradation of bagasse (Johnsrud et al. 1987; Johnsrud and Eriksson 1985).

The work of STFI is summarized in numerous publications on biomechanical pulping and related subjects (Ander and Eriksson 1975; Eriksson 1985; Eriksson et al. 1976, 1980; Eriksson and Vallander 1980, 1982; Johnsrud et al. 1987; Johnsrud and Eriksson 1985; Kirk 1993; Setliff et al. 1990). The wood handling system of an existing mill can easily be incorporated with the biopulping process, as shown in Fig. 5.6.

Sykes (1994) found that raw chip pulping effluents were significantly more toxic than fungus-treated wood chip mechanical pulping effluents (Table 5.12). Toxicity, as measured by the EPA, decreased from 33 to 4 (100/EC50). Additionally, fungal-treated chips' effluents from kraft pulping and sulfite pulping are less polluting and are less harmful than chips that have not been treated. It is common knowledge that waste loads rise when temperature rises, pulping chemicals are added, or pulp yields are reduced. As a result, biopulping's waste load should be significantly lower and less harmful than that of commercial mechanical or chemithermomechanical pulp mill effluents. Based on these findings, biopulping seems to be good for the environment.

Refining equipment	Energy input (kWh/ton)	
	CMP	BCMP
Defibrator and PFI mill	4800	1700
Disk refiner	3100	2100

Table 5.11 Energy requirement for chemimechanical pulp (CMP) and biochemimechanical pulp (BCMP) from bagasse

Based on data from Johnsrud et al. (1987)

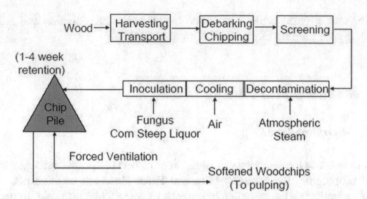

Fig. 5.6 Biopulping process can be fitted into an existing mill's wood handling system (Bajpai 2018b). Reproduced with permission

Table 5.12 BOD, COD, and toxicity of non-sterile aspen chips after treatment with *C. subvermispora*

Pulp[a]	COD (g/kg pulp)	BOD(g/kg pulp)	EPA toxicity[b] (100/EC$_{50}$)
Raw chips	40	18	33
Control			
No nutrients	30	10	5
Nutrients added	33	12	9
Fungus treated			
No nutrients	33	10	6
Nutrients added	35	11	4

[a] All pulps incubated for 4 weeks at 27 °C, except raw chips
[b] EC$_{50}$ is a measure of toxicity
Based on data from Sykes (1994)

Additionally, biopulping conserves a significant amount of electrical power. For commercial-scale applications, a biopulping procedure would frequently require inoculum, requiring additional effort and money.

It is typically challenging to produce basidiomycetes on a large scale. The treatment for the fungus is long. To reap the intended benefits, an incubation period of at least two weeks is required. The lengthy reaction time required by the fungal process appears to be a major drawback at first glance.

However, the pulp mill ought to have sufficient time and space to carry out this procedure because wood chips are frequently kept there for at least two weeks. However, traditional or molecular genetic approaches are preferred to shorten reaction treatment times and increase the biopulping fungi's effectiveness. Even though a pilot-scale biopulping system based on chip piles has been designed and tested, mill-scale operation must be demonstrated.

5.9.1.2 Enzymes

Enzymes have been also used to reduce specific energy consumption in refining of chips. The properties of the pulp, like its strength, opacity, and ability to scatter light, also improved as a result of the treatment (Pere et al. 2002). Between the primary refiner and the secondary refiner, enzymes are added to the wood chips. The enzymes make the cellulose fibers more free of fibers by hydrolyzing the hemicellulose. This would make it possible to shorten the amount of time needed in the secondary refiner.

In a variety of studies, enzymes such as laccase-mediator system, manganese peroxidase, pectinase, endoglucanase, and cellulase mixture were utilized to identify the most effective enzymes for lowering the amount of energy required for refinery operations. Cellulases act on the crystalline components of cellulose and create cavities in the structure of wood. This makes fibers more flexible and fibrillate. However, the yield gets also affected as cellulases destroy some of the cellulose (Aehle 2004; Hatakka et al. 2002; Pursula 2005).

Manganese peroxidases play a role in the processes of lignin degradation and have an effect on the structure of lignin. In procedures that involve the breakdown of plant materials, pectinase enzymes are frequently utilized. They can be utilized in the mechanical pulp manufacturing process because they are active between pH 4 and 5. When compared to the aggressive pectinase, the more selective less aggressive pectinase produced thermomechanical pulps with a higher scattering coefficient and higher pulp strength.

Pulps made with the enzyme blend of multiple components had the best pulp strength properties. For instance, in comparison to bisulfite pulp, pulps made with the pectinase pretreatment had the same pulp strength properties, a high scattering coefficient, and lesser generation of COD. The majority of enzymes tested for use in wood fiber treatments perform best between 40 and 55 °C (Kirk and Jeffries 1996; Michel 2003).

During production of mechanical pulp, the physical–chemical mechanism is not completely understood. It is unclear how various components of the fiber wall structure affect biopulping or the development of pulp properties particularly in industrial processes. Different fiber cell-wall components can be affected in a more controlled manner using biological methods. This broadens the range of resources available for production and opens up new possibilities for tailor-made TMP pulp.

5.9.2 Biochemical Pulping

5.9.2.1 Microorganisms

Pretreatment of wood with fungi has not received much attention prior to chemical pulping. However, a few studies examined pretreatments for kraft as well as sulfite pulping utilizing the fungi effective in biomechanical pulping (Bajpai et al. 2004, 2001; Messner et al. 1997).

Kraft

Compared to pulps made from untreated chips, kraft pulps made from aspen or
red oak chips that have been pretreated with *Phanerochaete chrysosporium* for 10 to
30 days cook more quickly and showed higher yield at a given kappa number (residual
lignin in pulp) (Oriaran et al. 1990, 1991).The fungus-treated chips improved cooking
properties and lower lignin content were attributed to increased cooking liquor penetration. According to Oriaran et al. (1990), the fungus-treated pulps had a higher
tensile strength and were more responsive to beating than the control pulps. These
studies did not address the environmental effects of the fungal treatments. However,
the reported yield increase would be negated by the significant (up to 17%) weight
loss of the wood during pretreatment with fungi, and the fungus's darkening of the
wood would probably necessitate the use of additional bleaching chemicals.

The active alkali dose in the cooking liquor reduced by 20% and the kappa number
by 29% after 21 days of pretreatment with the fungus Cartapip 97 (*O. piliferum*)
(Wall et al. 1996).Viscosity, which is a measure of pulp strength, increased while
pulp yield remained unchanged. The removal of pit membranes, resin deposits, and
ray parenchyma cells improved liquor penetration, resulting in the improvements.
The Cartapip pretreatment could be used to use less bleach chemicals both of which
have a negative impact on the environment (Wall et al. 1996).

Pycnoporus sanguineus, Stereum hirsutum and *Coriolus versicolor*, were found to
be the most promising of the 283 basidiomycetes examined in terms of their capacity
to make pine chips more efficient for kraft pulping (Wolfaardt et al. 1996). When
the pine chips were pretreated with *S. hirsutum* for a period of 9 weeks, the pulping
time to reach a kappa number of 28 reduced but consumption of alkali increased and
viscosity and pulp yield reduced. It didn't seem like the fungal pretreatment helped
the economy or the environment.

When eucalyptus chips were treated for two weeks with the fungus *C. subvermispora*, the extractive content decreased by 17–39% and the AA requirement decreased
by 18% (Bajpai et al. 2001). Biopulps were easily bleached and refined than the
control pulps, required lesser energy (18–30%), and had greater brightness and
strength. In another study by Bajpai et al. (2004), at the similar alkali charge, wheat
straw treated with the fungus *Ceriporiopsis subvermispora* for a week showed a
decrease in extractive content (44%) and a decrease in kappa number (22–27%). At
the same kappa value, it was discovered that the raw material had a 30 kg/t lower alkali
charge. Biopulps were brighter and whiter, required fewer bleaching chemicals, and
did not affect strength properties in any way.

Sulfite

In sulfite pulping, the rate of delignification of chips is accelerated when they
are pretreated with white rot fungi (Messner et al. 1997). When birch and spruce
chips were treated for two to four weeks with *Dichomitus squalens, Phlebia tremel-
losa, Phanerochaete brevispora*, and particularly *C. subvermispora*, the pulp kappa

number significantly decreased in magnesium-based sulfite pulping. However, the fungal treatments also diminished the brightness of the pulp. The pulp's strength properties were also affected. In sodium bisulfite pulping, pretreatment of loblolly pine chips with *C. subvermispora* for two weeks increased the rate of lignin loss and also the yield loss to the same extent, but it favoured delignification in calcium-acid sulfite pulping and reduced shives. The chips became darker as a result of the fungal treatment, necessitating the use of equal quantities of bleaching agents to lighten the reference and treated pulps. While the BOD and COD values in the reference and fungal treated effluents were identical, the microtox toxicity in the fungal treated chips' effluent was lesser than half in comparison to the control. According to Messner et al. (1997), the degradation of fatty acids and resin by fungus was responsible for the decreased toxicity. Delignification during the fungal treatment partially replaces delignification by the pulping chemicals in biochemical pulping, but there is no obvious benefit to the economy or the environment.

More promising are pretreatments using non-ligninolytic fungi or enzymes that are able to remove harmful extractives and/or obstructions to penetration of pulping liquor. Since there is less lignin present and chemical delignification presents more of a challenge, fungal delignification is more suitable.

5.9.2.2 Enzymes

Chips of maple, birch, sycamore, oak or pine were treated with an enzyme mixture that contained cellulases and hemicellulases for a period of 24 h. This resulted in an increase in the longitudinal and transverse diffusion rates of sodium hydroxide in the chips, blocked cell pores, and caused the pit membranes to rupture (Jacobs-Young et al. 1998). The anticipated increase in delignification for kraft pulping was supported by studies with enzyme-treated sycamore chips. When pectinase was added to the enzyme mixture, the kappa number decreased without any reduction in yield. The enzyme-treated chips' pulps were comparable in strength to the control pulps and were simpler to bleach with chlorine dioxide. Lower effluent BOD, COD, and chloroorganic loadings should result from a reduction in bleaching chemical use.

The white-rot fungus has been the focus of the majority of biopulping research on lignocelluloses. The laccase–mediator system has also been used to study biopulping (Vaheri et al. 1991; Dyer and Ragauskas 2004; Widsten and Kandelba 2008; Petit-Conil et al. 2002).

Pretreatment of softwood (pine) chips with laccase and HBT, VA or ABTS, before kraft pulping was conducted (Dyer and Ragauskas 2004). Laccase/HBT was found to be the most effective laccase mediator system, in terms of boosting pulp yield and delignification.

Petit-Conil et al. (2002) used laccases from three fungi and HBT to treat softwood (spruce) chips prior to TMP. With two of the laccases, laccase/HBT reduced refiner energy by up to 20%, while an increase was observed in case of the third laccase. The handsheets' mechanical strength and brightness were mostly improved as a result of the change in pulp properties. The fiber surface chemistry was changed, which led to an increase in the potential for external fibrillation and bonding in the pulp. When comparison was made to bleaching without laccase pretreatment, one of the laccases without mediator resulted in a 15% reduction in the amount of peroxide required for equal brightness. Vaheri et al. (1991) reported that during mechanical pulping, pretreatment with laccase cuts down on energy usage. The pulp gets stronger and shows a higher blue reflectance factor.

References

Abbott AP, Capper G, Davies DL, Rasheed RK, Tambyrajah V (2003) Novel solvent properties of choline chloride/urea mixtures. Chem Commun 1:70–71

Abbott AP, Boothby D, Capper G, Davies DL, Rasheed RK (2004) Deep eutectic solvents formed between choline chloride and carboxylic acids: versatile alternatives to ionic liquids. J Am Chem Soc 126(29):9142–9147

Abo-Hamad A, Hayyan M, AlSaadi MA, Hashim MA (2015) Potential applications of deep eutectic solvents in nanotechnology. Chem Eng J 273:551–567

Aehle W (2004) Enzymes in industry, production and application. Wiley-VCH, Weinheim, p 484

Agbor VB, Cicek N, Sparling R, Berlin A, Levin DB (2011) Biomass pretreatment: fundamentals toward application. Biotechnol Adv 29:675–685

Alejandro M, Saldivia G, Jan./Feb (2003) Two-Stage O2 Delignification System Cuts Mill's Chemical Use, Boosts Pulp Quality. PaperAge 18–24

Allison RW (1979) Effect of ozone on high-temperature thermo-mechanical pulp. Appita J 32(4):279–284

Allison RW (1980) Low energy pulping through ozone modification. Appita J 34(3):197–204

Alvarez-Vasco C, Ma R, Quintero M, Guo M, Geleynse S, Ramasamy KK, Wolcott M, Zhang X (2016) Unique low molecular-weight lignin with high purity extracted from wood by deep eutectic solvents (DES): a source of lignin for valorization. Green Chem 18:5133–5141

Ander P, Eriksson KE (1975) Influence of carbohydrates on lignin degradation by the white-rot fungus Sporotrichum pulverulentum. Sven Papperstid 78:643–652

Anttila JR, Rousu PP, Tanskanen JP (2006) Chemical recovery in nonwood pulping based on formic acid—application of reactive evaporation. In: New technologies in nonwood fiber pulping and papermaking: 5th international nonwood fiber pulping and papermaking conference. Guangzhou, China, pp 334–338

Asiz S, McDonough TJ (1987) Ester pulping. A brief evaluation. Tappi J 70(3):137–138

Asiz S, Sarkanen K (1989) Organosolv pulping. A review. Tappi J 72(3):169–175

Azeez MA (2018) Pulping of nonwoody biomass, pulp and paper processing, Salim Newaz Kazi, IntechOpen. https://doi.org/10.5772/intechopen, 79749. https://www.intechopen.com/books/pulp-and-paper-processing/pulpingof-nonwoody-biomass

Baeza J, Urizar S, Freer J, Rodríguez J, Peralta-Zamora P, Durán N (1999) Organosolv pulping. IX. Formic acid/acetone delignification of Pinus radiata and Eucalyptus globulus. Cellul Chem Technol 33(3/4):289–301

Baig K, Wu J, Turcotte G, Doan HD (2015) Novel ozonation technique to delignify wheat straw for biofuel production. Energy Environ 26:303–318

Bajpai P (2010) Environmentally friendly production of pulp and paper. Wiley, New York

Bajpai P (2018b) Biotechnology for pulp and paper processing. Springer Nature, Singapore

Bajpai P, Bajpai PK, Akhtar M (2001) Biokraft pulping of eucalyptus with selected lignin-degrading fungi. J Pulp Pap Sci 27(7):235–239

Bajpai P, Mishra SP, Mishra OP, Kumar S, Bajpai PK, Singh S (2004) Biochemical pulping of wheat straw. Tappi J 3(8):3–6

Bajpai P (2012) Environmentally benign approaches for pulp bleaching, 2nd ed. Elsevier B.V, 406 p

Bajpai P (2015) Minimum impact mill technologies. In: Green chemistry and sustainability in pulp and paper industry. Springer, Cham. https://doi.org/10.1007/978-3-319-18744-0_4

Bajpai P (2018a) Biermann's handbook of pulp and paper: volume 1: raw material and pulp making. Elsevier, USA

Bajpai P (2021) Nonwood plant fibers for pulp and paper. Chapter 7 pulping properties/pulping. Elsevier, pp 107–145

Balogh DT, Curvelo AAS (1998) Successive and batch extraction of Eucalyptus grandis in dioxane water-HCl solution. Paperi Ja Puu-Paper Timber 80(5):374–378

Binder A, Pelloni L, Fiechter A (1980) Delignification of straw with ozone to enhance biodegradability. Eur J Appl Microbiot Biotechnol 11:1–5

Bludworth J, Knopf FC (1994) Reactive extraction of lignin from wood using supercritical ammonia-water mixtures. J Supercrit Fluids 6(4):249–254

Bokström M, Nordén S (1998) Extended oxygen delignification. In: Proceedings of the 1998 international pulp bleaching conference. Helsinki, Finland

Brandt A, Grasvik J, Hallett J, Welton T (2013) Deconstruction of lignocellulosic biomass with ionic liquids. Green Chem 15:550–583

Bräuer P, Großalber J, Münster H, Zhang X, Nagalla RN (2012) China is steaming ahead with high-yield pulping, success story of Chinese paper and board industry with the use of mechanical pulping, what can Asia learn from this, http://papermart.in/2012/09/28/china-is-steaming-ahead-with-high-yield-pulping-success-story-of-chinese-paper-andboard-industry-with-the-use-of-mechanical-pulping-what-can-asia-learn-from-this

Cao Y, Li H, Zhang Y, Zhang J, He J (2010) Structure and properties of novel regenerated cellulose films prepared from cornhusk cellulose in room temperature ionic liquids. J Appl Polym Sci 116:547–554

Chaudhuri P (1996) Solvent pulping of bagasse. A process and system concept. In: TAPPI pulping conference proceeding, pp 583–594

Cherubini F (2010) The biorefinery concept: using biomass instead of oil for producing energy and chemicals. Energy Convers Manag 51:1412–1421

Chirat C, Lachenal D, Nyangiro D, Viardin MT, Janel K (2005) Applying ozone on high kappa pulps (kraft and sulfite) to improve the bleached pulp yield. EFPG DAYS 2005. Grenoble, France

Claus I, Kordsachia O, Schroeder N, Karstens T (2004) Monoethanolamine (MEA) pulping of beech and spruce wood for production for dissolving pulp. Holzforschung 5886:573–580

Colodette JL, Campos AS, Gomide JL (1990a) Attempts to use white liquor as the source of alkali comparison of chemical pretreatment methods for improving saccharification of cotton composition and degradation of wheat straw monosaccharides. Eur J Appl Concept Ind Crops Prod 108:431–441

Colodette JL, Campos AS, Gomide JL (1990b) Attempts to use white liquor as the source of alkali in the oxygen delignification of eucalypt kraft pulp. In: 1990 Tappi oxygen delignification symposium notes. Tappi Press, Atlanta, p 145

Constant S, Wienk HLJ, Frissen AE, Peinder PD, Boelens R, Es DSV, Grisel RJH, Weckhuysen BM, Huijgen WJJ, Gosselink RJA, Bruijnincx PCA (2016) New insights into the structure and composition of technical lignins: a comparative characterisation study. Green Chem 18(9):2651–2665

Dadi AP, Schall CA, Varanasi S (2007) Mitigation of cellulose recalcitrance to enzymatic hydrolysis by ionic liquid pretreatment. Appl Biochem Biotechnol 137–140(1–12):407–421

De Rosa MR, Da Silva C, Antonio A (1997) Organosolv delignification of wheat straw. In: 5th proceedings of Brazilian symposium on the chemistry of lignins and other wood components, vol 6, pp 224–231

De Santi V, Cardellini F, Brinchi L, Germani R (2012) Novel Bronsted acidic deep eutectic solvent as reaction media for esterification of carboxylic acid with alcohols. Tetrahedron Lett 53(38):5151–5155

Degam G (2017) Deep eutectic solvents synthesis, characterization and applications in pretreatment of lignocellulosic biomass. http://openprairie.sdstate.edu/etd (2017)

Delmas M, Benjelloun-Mlayah B, Avignon G (2006) Pilot plant production of pulp, linear lignin and xylose. In: Processing 60th Appita annual conference and exhibition. Melbourne, Australia, pp 283–287

Delpechbarrie F, Robert A (1993) Oxygen delignification in a water plus organic-solvent solution. 1. Delignification of poplar chips (Populus species) in a water-acetone solution. Cellul Chem Technol 27(1):87–105

Demirbas A (1998) Aqueous glycerol delignification of wood chips and ground wood. Biores Technol 63(2):179–185

Dillner B, Tibbling P (1991) Use of ozone at medium consistency for fully bleached pulp. Process concept and effluent characteristics. In: International pulp bleaching conference. Stockholm, June 11–14, Proceedings, vol 2, pp 59–74

Dyer TJ, Ragauskas AJ (2004) Laccase: a harbinger to kraft pulping. ACS Sym Ser 889:339–362

Elgharbawy AA, Hayyan M, Hayyan A, Basirun WJ, Salleh HM, Mirghani ME (2020) A grand avenue to integrate deep eutectic solvents into biomass processing. Biomass Bioenergy 137:105550

Elmasry AM, Mostafa NYS, Hassan HA, Aboustate MA (1998) Formamide and dimethylformamide and their effects on bagasse dissolving pulps. Cellul Chem Technol 32(5/6):433–440

Enqvist E, Tikka P, Heinrich L, Luhtanen M (2006) Production of pulp using a gaseous organic agent as heating and reaction accelerating media. Patent WO2006103317

Enz SM, Emmerling F (1987) North America's first fully integrated, medium consistency oxygen delignification stage. Tappi J 70(6):105–112

Eriksson KE (1985) Swedish developments in biotechnology related to the pulp and paper industry. Tappi J 68(7):46–55

Eriksson KE, Vallander L (1980) Biomechanical pulping. In: Kirk TK, Higuchi T, Chang H-M (eds) Lignin biodegradation: microbiology, chemistry, and potential applications, vol 2. CRC Press, USA, pp 213–233

Eriksson KE, Vallander L (1982) Properties of pulps from thermomechanical pulping of chips pretreated with fungi. Sven Papperstid 85:R33–R38

Eriksson KE, Grünewald A, Vallander L (1980) Studies of growth conditions in wood for three white-rot fungi and their cellulase less mutants. Biotech Bioeng 22:363–437

Eriksson KE, Ander P, Henningsson B (1976). Method for producing cellulose pulp. US Patent, 3,962,033

European Commission (2001) Integrated pollution prevention and control (IPPC). Reference document on best available techniques in the pulp and paper industry. Institute for Prospective Technological Studies, Seville

Feng L, Chen Z (2008) Research progress on dissolution and functional modification of cellulose in ionic liquids. J Mol Liq 142:1–5

Ferraz A, Rodríguez J, Freer J, Baeza J (2000) Formic acid/acetone-organosolv pulping of white rotted Pinus radiata softwood. J Chem Technol Biotechnol 75(12):1190–1196

Fink HP, Weigel P, Purz HJ (2001) Structure formation of regenerated cellulose materials from NMMO-solutions. J Prog Polym Sci 26:1473–1524

Ford M, Sharman P (1996) Performance of high yield hardwood pulp is investigated as it should be the choice of the future. Pulp Pap Int 38(10):29

Francisco M, van den Bruinhorst A, Kroon MC (2012) New natural and renewable low transition temperature mixtures (LTTMs): screening as solvents for lignocellulosic biomass processing. Green Chem 14(8):2153–2157

Freer J, Rodríguez J, Baeza J, Duran N, Urizar S (1999) Analysis of pulp and lignin extracted with formic acid-acetone mixture form Pinus radiata and Eucalyuptus globulus wood. Boletín De La Sociedad Chilena De Química 4482:199–207

Garcia-Cubero MT, Gonzalez-Benito G, Indacoechea I, Coca M, Bolado S (2009) Bioresource technology, effect of ozonolysis pretreatment on enzymatic digestibility of wheat and rye straw, vol 100, pp 1608–1613

Gast D, Puls J (1985) Ethylene glycol-water pulping. Kinetics of delignification. In: Ferrero GL (ed) Anaerobic digestion and carbohydrate hydrolysis of waste. Elsevier Applied Science Publishers Ltd., Essex, pp 450–453

Gorke JT, Srienc F, Kazlauskas RJ (2008) Hydrolase-catalyzed biotransformations in deep eutectic solvents. Chem Commun 10:1235–1237

Gullichsen J, Paulapuro H, Sundholm J (ed.) (2000) Papermaking science and technology, Book 5. Mechanical pulping, Fapet Oy, Helsinki, Finland

Gullichsen J (2000) Fiber line operations. In: Gullichsen J, Fogelholm C-J (eds) Chemical pulping—papermaking science and technology, Book 6A. Fapet Oy, Helsinki, Finland, p A19

Hadj-Kali M (2015) Separation of ethyl benzene and n-octane using deep eutectic solvents. Green Proc Synth 4(2):117–123

Han Y, Law KN, Lanouette R (2008). Modification of jack pine TMP long fibers by alkaline peroxide—Part 1. Chemical characteristics of fibers and spent liquor. BioResources 3(3):870–880

Hatakka A, Maijala P, Mettälä A (2002) Fungi as potential assisting agents in softwood pulping. Biotechnol Pulp Paper Ind 21:81–88

Heinze T, Schwikal K, Barthel S (2005) Ionic liquids as reaction medium in cellulose functionalization. Macromol Biosci 5:520–525

Hergert HL (1998) Developments in organosolv pulping. An overview. In: Young RA, Akhtar M (eds) Environmental friendly technologies for the pulp and paper industry. Wiley, New York

Holm J, Lassi U (2011). Ionic liquids in pretreatment of lignocellulosic biomass. In: Kokorin A (ed) Ionic liquids: application and perspectives. In-Tech, pp 546–560

Holmbom B, Ekman R, Sjoholm R, Eckerman C, Thornton J (1991) Chemical-changes in peroxide bleaching of mechanical pulps. Papier 45(10A):16–22

Hostachy JC (2010a) Ozone-enhanced bleaching of softwood kraft pulp. Tappi J 9(8):16–23

Hostachy JC (2010b) Softwood pulp bleaching with ozone: a new concept to reduce the bleaching chemical cost by 25%. Appita 63(2):92–97

Hostachy JC (2010c) Use of ozone in chemical and high yield pulping processes: latest innovations maximizing efficiency and environmental performance. In: 64th Appita annual conference and exhibition, 18–21 April 2010. Melbourne, Australia, pp 349–354

Hostachy JC (2010d) Use of ozone in chemical and high yield pulping processes: latest innovations maximizing efficiency and environmental performance. In: 64th Appita annual conference and exhibition, 18–21 April 2010. Melbourne, Australia, pp 349–354

Hou XD, Feng GJ, Ye M, Huang CM, Zhang Y (2017) Significantly enhanced enzymatic hydrolysis of rice straw via a high-performance two-stage deep eutectic solvents synergistic pretreatment. Bioresour Technol 238:139–146

Hultholm TEM, Nylund K, Lonnberg KB, Finell M (1995) The IDE-process: a new pulping concept for nonwood annual plants. Processing TAPPI pulping conference. Chicago, IL, Book 1, pp 85–89 (1995)

Isaifan RJ, Amhamed A (2018) Review on carbon dioxide absorption by choline chloride/urea deep eutectic solvents. Adv Chem 1–6

Jacobs-Young CJ, Venditti RA, Joyce TW (1998) Effect of enzymatic pretreatment on the diffusion of sodium hydroxide in wood. Tappi J 81(1):260–266

Jahan MS, Farouqui FI (2000) Pulping of whole jute plant (Corchorus capsularis) by soda-amine process. Holzforschung 54(6):625–630

Jahan MS, Farouqui FI (2003) Kinetics of jute pulping by soda-amine processes. Cellul Chem Technol 36(3/4):357–366

Jahan MS, Farouqui FI, Hasan AJM (2001) Kinetics of jute pulping by soda-amine process. Bangladesh J Sci Ind Res 36(1/4):25–31

Jahan MS (2001) Soda-amine pulping of cotton stalk. In: Pulping conference United States, pp 1175–1183

Jerschefske D (2012) China invests to meet booming paper demand. http://www.labelsandlabeling.com/news/features/china-invests-meet-booming-paper-demand

Jessop PG, Jessop DA, Fu D, Phan L (2012) Solvatochromic parameters for solvents of interest in green chemistry. Green Chem 14:1245–1259

Jiménez L, de la Torre MJ, Maestre F, Ferrer JL, Pérez I (1997a) Organosolv pulping of wheat straw by use of phenol. Biores Technol 60:199–205

Jiménez L, Maestre F, Pérez I (1997b) Disolventes orgánicos para la obtención de pastas con celulosa. Review. Afinidad 44(467):45–50

Jiménez L, de la Torre MJ, Bonilla JL, Ferrer JL (1998) Organosolv pulping of wheat straw by use of acetone-water mixtures. Process Biochem 33(4):401–408

Jiménez L, García JC, Pérez I, Ferrer JL, Chica A (2001b) Influence of the operating conditions in the acetone pulping of wheat straw on the properties of the resulting paper sheets. Biores Technol 79(1):23–27

Jiménez L, Villar JC, Rodríguez A, Jiménez RM, Calero A (2002a) Influence of pulping parameters of olive prunnings with ethanolamine and soda on pulp characteristics. Afinidad 59(500):399–408

Jiménez L, Domínguez JC, Pérez I (2003) Influence of cooking variables in the organosolv pulping of wheat straw using mixtures of ethanol-acetone and water. Tappi J 2(1):27–31

Jiménez L, Rodríguez A, Calero A, Eugenio ME (2004a) Use of ethanolamine-soda-water mixtures for pulping olive wood trimmings. Chem Eng Res Des 82(A8):1037–1042

Jiménez L, Rodríguez A, Pérez I, Calero A, Ferrer JL (2004b) Ethylene glycol-soda organosolv pulping of olive tree trimmings. Wood Fiber Sci 36(3):423–431

Jiménez L, García JC, Pérez I, Ariza J, López F (2001a) Acetone pulping of wheat straw. Influence of the cooking and beating conditions on the resulting paper sheets. Ind Eng Chem Res 40(26):6201–6206

Jiménez L, Pérez I, López F, Ariza J, Rodríguez A (2002b). Ethanol-acetone pulping of wheat straw. Influence of the cooking and the beating of the pulps on the properties of the resulting paper sheets. Bioresour Technol 83(2):139–143

Johnson RW, Bird A (1991) CTMP in fine papers: impact of CTMP on permanence of alkaline papers, 1991 papermakers conference proceedings. TAPPI Press, pp 267–273

Johnson AP, Johnson BI, Gleadow P, Silva FA, Aquilar RM, Hsiang CJ, Araneda H (2008) 21st century fibrelines. In: Proceedings of the international bleaching conference. Quebec City

Johnsrud SC, Eriksson KE (1985) Cross-breeding of selected and mutated homokaryotic strains of Phanerochaete chrysosporium K-3: new cellulase deficient strains with increased ability to degrade lignin. Appl Microbiol Biotechnol 21:320–327

Johnsrud SC, Fernandez N, Lopez P (1987) Properties of fungal pretreated high yield bagasse. Nordic Pulp Pap Res J 2:47–52

Keshavarzipour F, Tavakol H (2015) Deep eutectic solvent as a recyclable catalyst for three-component synthesis of β-amino carbonyls. Catal Lett 145(4):1062–1066

Khanolkar VD (1998) Punec pulping. Pudumjee develops clean nonwood pulping. Asia Pac Papermaker 8(12):32–33

Kirk TK, Jeffries TW (1996) Roles for microbial enzymes in pulp and paper processing. In: ACS Symposium, 13 pp

Kirk TK, Akhtar M, Blanchette RA (1994) Biopulping: seven years of consortia research. Processing tappi biology science symposia. Tappi Press, Atlanta, pp 57–66

Kirk TK (1993) Biopulping: a glimpse of the future? Forest products laboratory, madison, WI, Res Rep FPL-RP-523

Koell P, Lenhardt H (1987) Organosolv pulping of birch wood in a flow apparatus. Holzforschung 41(2):89–96

Kojima Y, Yoon SL (1991). Distribution of lignin in the cell wall of ozonized CTMP fibres. In: Proceedings from the 1991 6th international symposium on wood and pulp chemistry. Melbourne, Australia, p 109

Kubes GJ, Bolker HI (1978) Sulfur-free delignification. I. Alkaline pulping with monoethanolamine and ethylene diamine. Cellul Chem Technol 12(5):621–645

Kucuk MH, Demirbas A (1993) Delignification of Ailanthus altissima and Spruce orientalis with glycerol or alkaline glycerol at atmospheric pressure. Cellul Chem Technol 27(6):679–686

Kumar AK, Parikh BS, Pravakar M (2016) Natural deep eutectic solvent mediated pretreatment of rice straw: bioanalytical characterization of lignin extract and enzymatic hydrolysis of pretreated biomass residue. Environ Sci Pollut Res 23(10):9265–9275

Kunaver M, Anžlovar A, Žagar E (2016) The fast and effective isolation of nanocellulose from selected cellulosic feedstocks. Carbohyd Polym 148:251–258

Kuo CH, Lee CK (2009) Enhanced enzymatic hydrolysis of sugarcane bagasse by N-methylmorpholine-N-oxide pretreatment. Bioresour Technol 100:866–871

Lachenal D, Taverdet MT, Muguet M (1991) Improvement in the ozone bleaching of kraft pulps. International pulp bleaching conference. Stockholm, June 11–14, proceedings, vol 2, pp 33–43

Lecourt M, Struga B, Delagoutte T, Petit-Conil M (2007) Saving energy by application of ozone in the thermomechanical pulping process. IMPC, Minneapolis, pp 494–507

Lee JS (2017) Deep eutectic solvents as versatile media for the synthesis of noble metal nanomaterials. Nanotechnol Rev 6(3):271–278

Leponiemi A (2008) Nonwood pulping possibilities—a challenge for the chemical pulping industry. APPITA J 61(3):234–243

Leponiemi A (2011) Fibres and energy from wheat straw by simple practice. Doctoral dissertation, VTT Publication, p 767

Levlin JE (1990) On the use of chemi-mechanical pulps in fine papers. Pap Ja Puu-Pap Timber 72(4):301–308

Li K, Lei X, Lu L, Camm C (2010) Surface characterization and surface modification of mechanical pulp fibers. Pulp Pap Can 111(1):28–33

Li L, Yu ST, Liu FS, Xie CS, Xu CZ (2011) Efficient enzymatic in situ saccharification of cellulose in aqeous-ionic liquid media by microwave treatment. BioResources 6(4):4494–4504

Lindholm CA (1977a) Ozone treatment of mechanical pulp. Part 2: influence on strength properties. Pap Ja Puu 59(2):47–50, 53–58, 60, 62

Lindholm CA (1977b) Ozone treatment of mechanical pulp. Part 3: influence on optical properties. Pap Ja Puu 59(4a):217–218, 221–224, 227–232

Lindholm CA (1977c) Ozone treatment of mechanical pulps. Pap Ja Puu (Special No. 4a):217–231.

Liu LY, Chen HZ (2006) Enzymatic hydrolysis of cellulose materials treated with ionic liquid [BMIM]Cl. Chin Sci Bull 51:2432–2436

Loow Y-L, New EK, Yang GH, Ang LY, Foo LYW, Wu TY (2017) Potential use of deep eutectic solvents to facilitate lignocellulosic biomass utilization and conversion. Cellulose 24(9):3591–3618

Lynam JG, Kumar N, Wong MJ (2017) Deep eutectic solvents' ability to solubilize lignin, cellulose, and hemicellulose; thermal stability; and density. Bioresour Technol 238:684–689

Machado ASR, Sardinha RMA, Gomes de Acebedo E, Nunes da Ponte M (1994) High-pressure delignification of eucalyptus wood by 1,4-dioxane-carbon dioxane mixtures. J Supercrit Fluids 7(2):87–92

Magara K, Ikeda I, Tomimura Y, Hosoya S (1998) Accelerated degradation of cellulose in the presence of lignin during ozone bleaching. J Pulp Pap Sci 24(8):264

Mamleeva NA, Kharlanov AN, Kuznetsova MV, Kosyakov DS (2022) Delignification of wood of populus tremula by treatment with ozone. Russ J Phys Chem 96:2043–2052

Mansour OY, Selim IZ, Mohamed SA (1996) Physical characterization of pulps. I. Rice straw bleached by nonconventional multistage method and paper sheet making. Polym-Plast Technol Eng 35(4):567–580

Mbous YP, Hayyan M, Hayyan A, Wong WF, Hashim MA, Looi CY (2017 March–April) Applications of deep eutectic solvents in biotechnology and bioengineering-Promises and challenges. Biotechnol Adv 35(2):105–134

McDonough TJ (1996) Oxygen delignification. In: Dence CW, Reeve DW (eds) Pulp bleaching principles and practice. Tappi Press, Atlanta, p 213

Meighan BN, Lima DRS, Cardoso WJ, Baêta BEL, Adarme OFH, Santucci BS, Pimenta MTB, de Aquino SF, Gurgel LVA (2017) Two-stage fractionation of sugarcane bagasse by autohydrolysis and glycerol organosolv delignification in a lignocellulosic biorefinery concept. Ind Crops Prod 108:431–441

Menon V, Rao M (2012) Trends in bioconversion of lignocellulose: biofuels, platform chemicals and biorefinery concept. Prog Energy Combust Sci 38(4):522–550

Messner K, Koller K, Wall MB (1997) Fungal treatment of wood chips for chemical pulping. In: Young RA, Akhtar M (eds) Environmentally friendly technologies for the pulp and paper industry. Wiley, New York, pp 385–419

Michel PC (2003) Development of biotechnologies in the production of mechanical pulps (BioHYP). CTP, 26 pp

Miron JD, Ben-Ghedalia (1982) Effect of hydrolysing and oxidizing agents on the composition and degradation of wheat straw monosaccharaides. Eur J Appl Microbiol Biotechnol 15:83–87

Mondal D, Sharma M, Wang CH, Lin YC, Huang HC, Saha A (2016) Deep eutectic solvent promoted one step sustainable conversion of fresh seaweed biomass to functionalized grapheme as a potential electrocatalyst. Green Chem 18(9):2819–2826

Mora-Pale M, Meli L, Doherty TV, Linhardt RJ, Dordick JS (2011) Room temperature ionic liquids as emerging solvents for the pretreatment of lignocellulosic biomass. Biotechnol Bioeng 108(6):1229–1245

Mosier N, Hendrickson R, Brewer M, Ho N, Sedlak M, Dreshel R (2005a) Industrial scale-up of pH-controlled liquid hot water pretreatment of corn fiber for fuel ethanol production. Appl Biochem Biotechnol 125:77–97

Mosier N, Wyman CE, Dale BE, Elander R, Lee YY, Holtzapple MT (2005b) Features of promising technologies for pretreatment of lignocellulosic biomass. Bioresour Technol 96:673–686

Mostafa NYS (1994) Base-catalyzed dioxane and dioxane-borax pulping and fine structure, chemical reactivity and viscose filterability of cotton cellulose. Cellul Chem Technol 28(2):171–175

Moultrop JS, Swatloski RP, Moyna G, Rogers RD (2005) High resolution 13C NMR studies of cellulose and cellulose oligomers in ionic liquid solutions. R Soc Chem Chem Commun 12:1557–1559

Muhammad N, Man Z, Bustam MA, Mutalib MIA, Wilfred CD, Rafiq S (2011) Dissolution and delignification of bamboo biomass using amino acid-based ionic liquid. Appl Biochem Biotechnol 165(3–4):998–1009

Musale RM, Shukla SR (2016) Deep eutectic solvent as effective catalyst for aminolysis of polyethylene terephthalate (PET) waste. Int J Plast Technol 20(1):106–120

Muurinen E, Jurva E, Lahtinen I, Sohlo J (1993) Peroxyacid pulping and recovery. In: 7th international symposium on wood and pulping chemistry. Beijing, China, pp 195–200

Muurinen E (2000a) Organosolv pulping. Academic dissertation, Faculty of Technology, University of Oulu, Linnanmaa. Ministry of Food, Agriculture and Forestry

Muurinen E (2000b) Organosolv pulping. A review and distillation study related to peroxyacid pulping. Tesis doctoral. Departamento de Ingeniería de Procesos, Universidad. de Oulu, Finlandia

Neto PG, Delpechbarrie F, Robert A (1993) Oxygen delignification in a water plus organic solvent solution. 2. Comparison of eucalyptus wood (Eucalyptus globulus) and poplar wood (Populus species). Cellul Chem Technol 27(2):185–199

Nimz HH (1989) Pulping and bleaching by the Acetosolv process. Papier 43 (10A), V102–V108

Ninomiya K, Yamauchi T, Kobayashi M (2013) Cholinium carboxylate ionic liquids for pretreatment of lignocellulosic materials to enhance subsequent enzymatic saccharification. Biochem Eng J 71:25–29

Obst JR, Sanyer N (1980) Effect of quinones and amines on the cleavage rate of β-O-4 ethers in lignin during alkaline pulping. Tappi J 63(7):111–114

Oliet M (1999) Estudio sobre la deslignificación de eucalyptus globulus con etanol/agua como medio de cocción. [Tesis doctoral]. Madrid, Spain: Departamento de Ingeniería Química, Universidad Complutense de Madrid

Oliveira VKD, Gregory C, Francois J (2015) Contribution of deep eutectic solvents for biomass processing: opportunities, challenges, and limitations. ChemCatChem 7(8):1250–1260

Oriaran TP, Labosky P Jr, Blankenhorn PR (1990) Kraft pulp and papermaking properties of Phanerochaete chrysosporium degraded aspen. Tappi J 73(7):147–152

Oriaran TP, Labosky P Jr, Blankenhorn PR (1991) Kraft Pulp and papermaking properties of Phanerochaete chrysosporium degraded red oak. Wood Fiber Sci 23:316–327

van Osch DJGP, Kollau LJBM, van den Bruinhorst A, Asikainen S, Alves da Rocha MA, Kroon MC (2016) Ionic liquids and deep eutectic solvents for lignocellulosic biomass fractionation. Phys Chem Chem Phys 19(4):2636–2665

Paiva A, Craveiro R, Aroso I, Martins M, Reis RL, Duarte ARC (2014) Natural deep eutectic solvents–solvents for the 21st century. ACS Sustain Chem Eng 2(5):1063–1071

Pan GX (2001) An insight into the behaviour of aspen CTMP in peroxide bleaching—alkalinity's influence is greater than that of peroxide charge. Pulp Pap Can 102(11):41–45

Papa G, Varanasi P, Sun L (2012) Exploring the effect of different plant lignin content and composition on ionic liquid pretreatment efficiency and enzymatic saccharification of Eucalyptus globulus L. Mutants. Bioresour Technol 117:352–359

Patil UB, Singh AS, Nagarkar JM (2014) Choline chloride based eutectic solvent: an efficient and reusable solvent system for the synthesis of primary amides from aldehydes and from nitriles. RSC Adv 4(3):1102–1106

Patt R, Kordsachia O (1986) Production of pulps using alkaline sulphite solutions with the addition of anthraquinone and methanol. Papier 40(10a):V1–V8

Patt R, Kordsachia O, Reuter G (1987) Dtsch. Papierwirtschaft 3, T96–T102

Patt R, Kordsachia O, Shackford LD, Rockvam LN (1999) Conversion of an acid sulfite mill to the ASAM process for improved quality and economics. In: Processing TAPPI pulping conference, vol 2. Orlando, Florida, pp 667–676

Pere J, Ellmen J, Honkasalo J, Taipalus P (2002) Enhancement of TMP reject refining by enzymatic modification of pulp carbohydrates-A mill study. Biotechnol Pulp Paper Ind 21:281–290

Perez-Pimienta JA, Lopez-Ortega MG, Varanasi P (2013) Comparison of the impact of ionic liquid pretreatment on recalcitrance of agave bagasse and switchgrass. Bioresour Technol 127:18–24

Petit-Conil M, Semar S, Niku-Paavola M-L, Sigoillot JC, Asther M, Anke H (2002) Potential of laccases in softwood-hardwood high-yield pulping and bleaching. Prog Biotechnol 21:61–71

Petit-Conil M, de Choudens C, Espilit T (1998) Ozone in the production of softwood and hardwood high-yield pulps to save energy and improve quality. Nord Pulp Pap Res J 13(1):16–22

Pikka O, Vessala R, Vilpponen A, Dahllof H, Germgard U, Norden S, et al. (2000) Bleaching Applications. In: Gullichsen, J, Fogelholm, C.-J. (Eds.), Chemical Pulping—Papermaking Science and Technology. Fapet Oy, Helsinki, Finland: Book 6A, p. A19

Poppius-Levlin K (1991) Milox pulping with acetic acid peroxyacetic acid. Pap Puu 73(2):154–158

Procentese A, Johnson E, Orr V, Garruto Campanile A, Wood JA, Marzocchella A, Rehmann L (2015) Deep eutectic solvent pretreatment and subsequent saccharification of corncob. Bioresour Technol 192:31–36

Procentese A, Raganati F, Olivieri G, Russo ME, Rehmann L, Marzocchella A (2018) Deep eutectic solvents pretreatment of agro-industrial food waste. Biotechnol Biofuels 11(1):37

Pu YQ, Jiang N, Ragauskas AJ (2007) Ionic liquids as a green solvent for lignin. J Wood Chem Technol 27:23–33

Pursula T (2005) Bringing life to paper, biotechnology in the forest industry. KCL research project, pp 1136–201

Pye EK, Lora JH (1991) The Alcell process. In: Papex '91: PITA annual conference. Manchester, UK, 7 p

Ragauskas AJ, Beckham GT, Biddy MJ, Chandra R, Chen F, Davis MF, Davison BH, Dixon RA, Gilna P, Keller M (2014) Lignin valorization: improving lignin processing in the biorefinery. Science 344:1246843

Reis R (2001) The increased use of hardwood high yield pulps for functional advantages in papermaking. In: Proceedings of the 2001 papermakers conference. Cincinnati, OH, USA, pp 87–108

Ren H, Chen C, Wang Q, Zhao D, Guo S (2016) The properties of choline chloride based deep eutectic solvents and their performance in the dissolution of cellulose. BioResources 11(2):5435–5451

Rezati-Charani P, Mohammadi-Rovshandeh J (2005) Effect of pulping variable with dimethyl formamide on the characteristics of bagasse-fiber. Bioresour Technol 96(15):1658–1669

Rezayati-Charani P, Mohammadi-Rovshandeh J, Hashemi SJ, Kazemi-Najafi S (2006) Influence of dimethyl formamide pulping of bagasse on pulp properties. Biores Technol 97(18):2435–2442

Robert DR, Szadeczki M, Lachenal D (1999) Chemical characteristics of lignins extracted from softwood TMP after O_3 and ClO_2 treatment. In: Proceedings from the 215th national ACS meeting, Lignin: historical, biological and materials perspectives, Dallas, Texas, USA, chapter 27, pp 520–531

Rodríguez A, Serrano L, Moral A, Pérez A, Jiménez L (2008) Use of high-boiling point organic solvents for pulping oil palm empty fruit bunches. Biores Technol 99(6):1743–1749

Rousu PP, Rousu P (2000) Method of producing pulp using single-stage cooking with formic acid and washing with performic acid. US patent 6156156

Rousu P, Rousu P, Anttila J (2002) Sustainable pulp production from agricultural waste. Resour Conserv Recycl 35(1–2):85–103

Rousu PP, Rousu P, Rousu E (2003) Process for producing pulp with a mixture of formic acid and acetic acid as cooking chemical. US patent 6562191

Roy-Arcand L, Archibald F (1996) Selective removal of resin and fatty acids from mechanical pulp effluents by ozone. Water Res 30(5):1269–1279

Ruffini G (1966) Improvement of bonding in wood pulps by the presence of acidic groups. Svensk Papperstding 69(3):72

Rutkowski J, Mroz W, Perlinskasipa K (1995) Glycol-acetic wood delignification. Cellul Chem Technol 28(6):621–628

Rutkowski J, Mroz W, Surna-Slusarsaka B, Perlinskasipa K (1993) Glycolic delignification of hardwood. In: Progress 93 conference proceeding, vol 1, pp 190–205

Saake B, Lehnen R, Lummitsch S, Nimz HH (1995) Production of dissolving and paper grade pulps by the formacell process. In: 8th international symposium on wood and pulping chemistry. Helsinki, Finland, pp 237–242

Salehi K, Kordsachia O, Patt R (2015) Comparison of MEA/AQ, soda and soda/AQ pulping of wheat and rye straw. Ind Crops Prod 52:603–610

Sano Y, Endo M, Sakashta Y (1989) Solvolysis pulping of softwoods with aqueous cresols containing a small amount of acetic acid. Mokuzai Gakkaishi 35(9):807–812

Sano Y, Shimamoto S (1995) Pulping of birchwood at atmospheric pressure with aqueous acetic acid containing small amounts of sulfuric acid and phenols. Mokuzai Gakkaishi 41(11):1006–1011

Sarwar MJ, Farouqui FI, Abdullah IS (2002) Pulping of jute with amines. Cellul Chem Technol 35(1/2):177–187

Sathitsuksanoh N, Zhu Z, Zhang Y-HP (2012) Cellulose solvent and organic solvent-based lignocellulose fractionation enabled efficient sugar release from a variety of lignocellulosic feedstocks. Bioresource Technol 117:228–233

Satlewal A, Agrawal R, Bhagia S, Sangoro J, Ragauskas AJ (2018) Natural deep eutectic solvents for lignocellulosic biomass pretreatment: recent developments, challenges and novel opportunities. Biotechnol Adv 36(8):2032–2050

Savcor Indufor (2007) Pulp quality comparison, Technical report—module 6, 13 p. http://www.ktm.fi/files/17224/Module_6_Final.pdf

Schroeter MC, Dahlmann G (1991) Organocell simplifies the solvent sulping process. In: Processing TAPPI pulping conference. Orlando, FL, pp 645–652

Schweers WHM (1974) Phenol pulping. Chem Technol 4(8):490–493

Schweers W, Behler H, Beinhoff O (1972) Pulping of wood with phenols II. Phenol balance. Holzforschung 26(3):103–105

Selim IZ, Mansour OY, Mohamed SA (1996) Physical characterization of pulps. II. Rice straw and bagasse pulps bleached by monoconventional two-stage hydrogen peroxide method and paper sheet making. Polym-Plast Technol Eng 35(5):649–667

Setliff EC, Marton R, Granzow SG (1990) Biomechanical pulping with white-rot fungi. Tappi J 73(8):141–147

Shishov A, Bulatov A, Locatelli M, Carradori S, Andruch V (2017) Application of deep eutectic solvents in analytical chemistry. A Rev Microchemical J 135:33–38

Silverstein RA, Chen Y, Sharma-Shivappa RR, Boyette MD, Osborne J (2007) A comparison of chemical pretreatment methods for improving saccharification of cotton stalks. Bioresour Technol 98:3000–3011

Simões R, Castro JA (1999) Ozone delignification of pine and eucalyptus kraft pulps. 2. Selectivity. Ind Eng Chem Res 38:4608–4614

Singh B, Lobo H, Shankarling G (2011) Selective N-alkylation of aromatic primary amines catalyzed by bio-catalyst or deep eutectic solvent. Catal Lett 141(1):178–182

Smith EL, Abbott AP, Ryder KS (2014) Deep eutectic solvents (DESs) and their applications. Chem Rev 114(21):11060–11082

Socha AM, Parthasarathi R, Shi J, Pattathil S, Whyte D, Bergeron M, George A, Tran K, Stavila V, Venkatachalam S, Hahn MG, Simmons BA, Singh S (2014) Efficient biomass pretreatment using ionic liquids derived from lignin and hemicellulose. PNAS. https://doi.org/10.1073/pnas.1405685111

Soteland N (1982) Interstage ozone treatment of hardwood high yield pulp. Pap Ja Puu 64(11):707–708, 710, 712–714

Spronsen JV, Witkamp GJ, Hollmann F, Choi YH, Verpoorte R (2011) Process for extracting materials from biological material. Patent: WO 2011155829 (A1) European Patent Office

Sun Y, Cheng J (2002) Hydrolysis of lignocellulosic materials for ethanol production: a review. Bioresour Technol 83:1–11

Sun Y, Lanouette R, Cloutier JN, Pelletier E, Épiney M (2014b) Impact of selective refining combined with inter-stage ozone treatment on thermomechanical pulp. BioResources 9(1):1225–1235

Sun Y, Lanouette R, Pelletier E, Cloutier JN, Epiney M (2013) Impact of pH during an interstage ozone treatment of thermomechanical pulp. In: PACWEST conference. Kamloops, BC, Canada, 6 p

Sun Y, Lanouette R, Pelletier E, Cloutier JN, Epiney M (2014a). Fibre performance of mechanical pulp after selective refining combined with interstage ozone treatment. In: IMPC conference. Helsinki, Finland, 10 p

Surma-Slusarska B (1998) Balance of ethylene glycol in organosolv pulping of hardwood. Przeglad Paper 54(12):712–714

Sykes M (1994) Environmental compatibility of effluents of aspen biomechanical pulps. Tappi J 77(1):160–166

Tang B, Row KH (2013) Recent developments in deep eutectic solvents in chemical sciences. Monatshefte Fur Chemie 144(10):1427–1454

Tang X, Zuo M, Li Z, Liu H, Xiong C, Zeng X, Sun Y, Hu L, Liu S, Lei T, Lin L (2017) Green processing of lignocellulosic biomass and its derivatives in deep eutectic solvents. Chemsuschem 10(13):2696–2706

Tatsuishi H, Hatano T, Iwai T, Kovasin K (1987) Practical experiences of medium consistency oxygen delignification by Rauma-Repola and Sumitomo heavy industries. In: Tappi international oxygen delignification conference proceedings. Tappi Press, Atlanta, p 209

Tench L, Harper S (1987) Oxygen bleaching practices and benefits – an overview. In: Tappi international oxygen delignification conference proceedings. Tappi Press, Atlanta, p 1

Tran PH, Nguyen HT, Hansen PE, Le TN (2016) An efficient and green method for regio- and chemo-selective friedel-crafts acylations using a deep eutectic solvent ([CholineCl][ZnCl$_2$]$_3$). RSC Adv 6:37031–37038

Ünlü AE, Takaç S (2020) Use of deep eutectic solvents in the treatment of agro-industrial ligno-cellulosic wastes for bioactive compounds [Online First]. IntechOpen, https://doi.org/10.5772/intechopen.92747

Usta M, Eroglu H, Karaoglu C (1999) ASAE pulping of wheat straw (Triticum aestivum L.). Cellul Chem Technol 33(1–2):91–102

Vaheri M, Salama N, Ruohoniemi K (1991) Procedure for the production of pulp. Eur Pat Appl EP429422. 29 May 1991

Vaidya AA, Murton KD, Smith DA, Dedual G (2022) A review on organosolv pretreatment of softwood with a focus on enzymatic hydrolysis of cellulose. Biomass Conv Bioref 12:5427–5442

Vancov T, Alston AS, Brown T, McIntosh S (2012) Use of ionic liquids in converting lignocellulosic material to biofuels. Renew Energy 45:1–6

Vasudevan B, Panchapakesan B, Gratzl, JS, Holmbom B (1987) The effect of ozone on strength development and brightness reversion characteristics of high yield pulps. In: Proceedings from the 1987 tappi pulping conference, proceedings. Washington, DC, pp 517–523

Vega A, Bao M, Lamas J (1997) Application of factorial design to the modelling of organosolv delignification of miscanthus sinensis (elephant grass) with phenol and dilute acid solutions. Biores Technol 61:1–7

Vega A, Bao M (1993) Organosolv fractionation of Ulex europaeus with dilute hydrochloric acid and phenol. Two simple kinetic models for prehydrolisis and delignification. Wood Sci Technol 27(1):61–68

Ventura SPM, Santos LDF, Saraiva JA, Coutinho JAP (2012) Concentration effect of hydrophilic ionic liquids on the enzymatic activity of Candida antarctica lipase B. World J Microbiol Biotechnol 28:2303–2310

Vidal PF, Molinier J (1988) Ozonolysis of lignin–improvement of in vitro digestibility of poplar sawdust. Biomass 161–167

Vigier KDO, Chatel G, Jerome F (2015) Contribution of deep eutectic solvents for biomass processing: opportunities, challenges, and limitations. ChemCatChem 7:1250–1260

Wall MB, Stafford G, Noel Y (1996) Treatment with Ophiostoma piliferum improves chemical pulping efficiency. In: Srebotnik E, Messner K (eds) Biotechnology in the pulp and paper industry. Recent advances in applied and fundamental research. Facultas-Universitatsverlag, Vienna, Austria, pp 205–210

Wallis AFA (1978) Wood pulping with mono-, di- and triethanolamine. Appita J 31(6):443–448

Wallis AFA (1980) Wood pulping with monoethanolamine in pressure vessels. Appita J 33(5):351–355

Wang K, Yang HY, Xu F, Sun RC (2011) Structural comparison and enhanced enzymatic hydrolysis of the cellulosic preparation from populus tomentosa Carr. by different cellulose-soluble solvent systems. Bioresour Technol 102:4524–4529

Wang L, Zhou M, Chen Q, He MY (2013) Bronsted acidic deep eutectic solvent catalysed the one-pot synthesis of 2H-indazolo[2,1-b]phthalazinetriones. J Chem Res 37(10):598–600

Wasserscheid P, Keim W (2000) Ionic liquids—new solutions for transition metal catalyst. Angew Chem Int Ed 39:3773–3789

Weerachanchai P, Leong SSJL, Chang MW, Ching CB, Lee JM (2012) Improvement of biomass properties by pretreatment with ionic liquids for bioconversion process. Bioresour Technol 111:453–459

Weil JR, Sariyaka A, Rau SL, Goetz J, Ladisch CM, Brewer M (1997) Pretreatment of yellow poplar wood sawdust by pressure cooking in water. Appl Biochem Biotechnol 68:21–40

Westmoreland RA, Jefcoat IA (1991) Sulfur dioxide-ethanol-water pulping of hardwoods. Chem Eng Commun 104:101–115

Widsten P, Kandelba A (2008) Laccase applications in the forest products industry: a review. Enzyme Microbial Technol 42(2008):293–307

Winner SR, Minogue LA, Lora JH (1997) ALCELL pulping of annual fibers. In: 9th international symposium on wood and pulping chemistry, Poster presentations. pp 120–1–120–4

Wolfaardt JF, Bosman JL, Jacobs A (1996) Bio-kraft pulping of softwood. Biotechnology in the pulp and paper industry. In: Srebotnik E, Messner K (eds) Recent advances in applied and fundamental research. Facultas-Universitatsverlag, Vienna, Austria, pp 211–216

Wright JD (1998) Ethanol from biomass by enzymatic hydrolysis. Chem Eng Prog 84(8):62–74

Wu J, Zhang J, He J (2004) Homogeneous acetylation of cellulose in a new ionic liquid. Biomacromol 5:266–268

Wu H, Mora-Pale M, Miao J, Doherty TV, Linhardt RJ, Dordick JS (2011) Facile pretreatment of lignocellulosic biomass at high loadings in room temperature ionic liquids. Biotechnol Bioeng 108(12):2865–2875

Xie RQ, Li XY, Zhang YF (2012) Cellulose pretreatment with 1-methyl-3-methylimidazoliumdimethylphosphate for enzymatic hydrolysis. Cellul Chem Technol 46(5–6):349–356

Xu EC (2001) P-RC alkaline peroxide mechanical pulping of hardwood, part 1: aspen, beech, birch, cottonwood and maple. Pulp Paper Can 102(2):44–47

Xu GC, Ding JC, Han RZ, Dong JJ, Ni Y (2016) Enhancing cellulose accessibility of corn stover by deep eutectic solvent pretreatment for butanol fermentation. Bioresour Technol 203:364–369

Xu P, Zheng GW, Zong MH, Li N, Lou WY (2017) Recent progress on deep eutectic solvents in biocatalysis. Bioresour Bioprocess 4(1):34

Yanhong G, Jing S, Qun L (2015) China's high-yield pulp sector and its carbon dioxide emission: considering the saved standing wood as an increase of carbon storage. BioResources 10(1):10–13

Yoo CG, Pu Y, Ragauskas AJ (2017) Ionic liquids: promising green solvents for lignocellulosic biomass utilization. Curr Opin Green Sustain Chem 5:5–11

Young RA, Akhtar M (1998) Environmentally friendly technologies for the pulp and paper industry. Wiley, New York, ISBN 0-471-15770-8

Young RA, Baierl KW (1985) Ester pulping of wood: a revolutionary process. Southern Pulp Paper 48:15–17

Zainal-Abidin MH, Hayyan M, Hayyan A, Jayakumar NS (2017 Aug 1) New horizons in the extraction of bioactive compounds using deep eutectic solvents: a review. Anal Chim Acta 979:1–23

Zavrel M, Bross D, Funke M, Buchs J, Spiess AC (2009) High-throughput screening for ionic liquids dissolving (ligno-) cellulose. Bioresour Technol 100:2580–2587

Zhang Z, O'Hara IM, Doherty WOS (2012a) Pretreatment of sugarcane bagasse by acid-catalysed process in aqueous ionic liquid solutions. Bioresour Technol 120:149–156

Zhang T, Kumar R, Wyman CE (2013a) Sugar yields from dilute oxalic acid pretreatment of maple wood compared to those with other dilute acids and hot water. Carbohydr Polym 92:334–344

Zhang DS, Yang Q, Zhu JY, Pan XJ (2013b) Sulfite (SPORL) pretreatment of switchgrass for enzymatic saccharification. Bioresour Technol 129:127–134

Zhang CW, Xia SQ, Ma PS (2016a) Facile pretreatment of lignocellulosic biomass using deep eutectic solvents. Biores Technol 219:1–5

Zhang K, Pei Z, Wang D (2016b) Organic solvent pretreatment of lignocellulosic biomass for biofuels and biochemicals: a review. Biores Technol 199:21–33

Zhang Q, De Oliveira VK, Royer S, Jerome F (2012b) Deep eutectic solvents: syntheses, properties and applications. Chem Soc Rev 41:7109–7146

Zhi S, Yu X, Wang X, Lu X (2012) Enzymatic hydrolysis of cellulose after pretreated by ionic liquids: focus on one-pot process. Energy Procedia 14:1741–1747

Zhou Y, Zhang D, Li G (2005) An overview of BCTMP: process, development, pulp quality and utilization. China Pulp Pap 24(5):51–60

Zhou Y (2004) Overview of high yield pulps (HYP) in paper and board. In: PAPTAC 90th annual meeting. Montreal, Canada, pp B143–B148

Zhu SD (2008) Perspective used of ionic liquids for the efficient utilization of lignocellulosic materials. J Chem Technol Biotechnol 83:777–779

Zulkefli S, Abdulmalek E, Abdul Rahman MB (2017) Pretreatment of oil palm trunk in deep eutectic solvent and optimization of enzymatic hydrolysis of pretreated oil palm trunk. Renewable Energy, Elsevier, vol 107(C):36–41

Index

A

ABTS, 73
Acetate, 10, 61
Acetic acid, 38, 49, 51–53, 55–59
Acetocell process, 58
Acetone, 50, 51, 58, 59, 63
Acetosolv method, 58
Acid bisulfite, 12
Acid catalyst, 38
Acidification, 23, 56
Acid rain, 2
Acute hemolytic anaemia, 33
Additives, 17, 47
Adhesives, 13, 18, 51
Aerobic, 13
Agricultural waste, 7
Air emissions, 23, 25, 26, 31, 34
Air pollution, 2, 31
ALCELL process, 55
Aldonic acids, 42
Alkaline peroxide mechanical pulp
    (APMP), 39, 40
Alkaline sulfite, 12, 56
Alkaline sulfite anthraquinone methanol
    (ASAM), 52, 54, 55, 57
Alkaline sulfite AQ-methanol pulping, 56
Alkaline sulfite liquor, 54
1-allyl-3-methylimidazolium chloride
    (Amim-Cl), 61, 62
Amines, 51, 59, 60
Ammonia, 59, 60
Ammonium sulfide, 59
Anaerobic, 13
Anthraquinone, 4, 52, 54–56
Areosols, 2
ASAE process, 55

ASAM process, 54, 55
Autohydrolysis, 38
Auxiliary boiler, 31

B

Bagasse, 7, 8, 52, 57–60, 62, 63, 69
Bamboo, 7, 61, 62
Bark, 9, 27
Bark furnace, 31
Basidiomycetes, 70, 72
BAT, 2
Battelle-Genoa method, 53
Beech, 52, 60
Bending force, 17
Benthic invertebrates, 27
Bioaccumulation, 23
Biochemical pulping, 71, 73
Biofuels, 49, 67
Biogas, 56
Biomass delignification, 49
Biomass pretreatment, 63, 67
Biomechanical pulping, 69, 71
Biopulp, 72
Biopulping, 37, 68–71, 73
Biotechnological method, 4
Biotechnological process, 4
Birch, 42, 52, 72, 73
Bisulfite, 12, 71, 73
Black liquor, 14, 32, 53
Black liquor oxidation, 14
Bleached chemi-thermomechanical pulp
    (BCTMP), 39, 40
Bleached pulp mill
    printing and writing papers, 10

P. Bajpai, *Environmentally Benign Pulping*, SpringerBriefs in Green Chemistry for Sustainability, https://doi.org/10.1007/978-3-031-23693-8

Printed in the United States
by Baker & Taylor Publisher Services